BIRDWATCHING IN NEW YORK CITY

AND ON LONG ISLAND

Deborah Rivel & Kellye Rosenheim

Kellye Rosenheim

Birdwatching

IN NEW YORK CITY
AND ON LONG ISLAND

Melissa & Larry,

My favorite Gipsy

birders

Love, Kellye

University Press of New England
Hanover and London

University Press of New England
www.upne.com
© 2016 Deborah Rivel and Kellye Rosenheim
All rights reserved
Manufactured in the United States of America
Designed by Mindy Basinger Hill
Typeset in Calluna

For permission to reproduce any of the
material in this book, contact Permissions,
University Press of New England,
One Court Street, Suite 250, Lebanon NH
03766; or visit www.upne.com

All images courtesy of Deborah Rivel, © 2016

Library of Congress Cataloging-in-Publication
Data

Names: Rivel, Deborah. | Rosenheim, Kellye.

Title: Birdwatching in New York City
and on Long Island / Deborah Rivel and
Kellye Rosenheim.

Other titles: Bird watching in New York City
and on Long Island

Description: Hanover: University Press of
New England, [2016] | Includes bibliographical
references and index.

Identifiers: LCCN 2015042509 (print) |
LCCN 2016004597 (ebook) | ISBN 9781611686784
(pbk.) | ISBN 9781611689686 (epub, mobi & pdf)

Subjects: LCSH: Bird watching—New York
Metropolitan Area. | Bird watching—New
York (State)—Long Island. | Birds—New York
Metropolitan Area—Identification. | Birds—
New York (State)—Long Island—Identification.

Classification: LCC QL677.5.R58 2016 (print) |
LCC QL677.5 (ebook) |
DDC 598.072/347471—dc23

LC record available at http://lccn.loc
.gov/2015042509

5   4   3   2   1

TO BOB GOODALE
*without whose vision and
faith in me this book
and so many other things
wouldn't have happened.*
DEB

**KELLYE
THANKS
JEFF, PHOEBE,
AND JULIA**
*whose love
supports her
in all things.*

Contents

Acknowledgments xv

INTRODUCTION I

1 MANHATTAN 17

Key Sites 19

Central Park 19
Inwood Hill Park 29
Fort Tryon Park 35
Sherman Creek and Swindler Cove 36
Randall's Island 38
Governors Island 43
Hudson River Greenway Biking 45
The Battery 48

Other Places to Find Birds in Manhattan 49

Bryant Park 50
Madison Square Park 50
Union Square Park 50
Washington Square Park 51
Morningside Park 52
Riverside Park and "the Drip" 52
Carl Schurz Park and the East River 53
Peter Detmold Park 53

Uniquely Manhattan Birding 53

2 BROOKLYN 57

Key Sites 57

Prospect Park 57
Brooklyn Botanic Garden 62
Green-Wood Cemetery 65
Floyd Bennett Field 69
Dead Horse Bay and Dead Horse Point 72

Coastal Brooklyn Winter Waterfowl Viewing 73

Brooklyn Bridge Park 74
Bush Terminal Piers Park 74
Brooklyn Army Terminal Pier 4 75
Owls Head Park and American Veterans Memorial Pier 75
Gravesend Bay 75
Calvert Vaux Park 76
Coney Island Creek 76
Coney Island Creek Park 77
Coney Island Pier 77
Plumb Beach 77
Salt Marsh Nature Center at Marine Park 79
Dead Horse Bay and Dead Horse Point 79
Canarsie Pier 80
Canarsie Park 80
Fresh Creek Park 81
Hendrix Creek and Betts Creek 81
Spring Creek Park 81

Other Places to Find Birds in Brooklyn 82

Brooklyn Bridge Park 82
Bush Terminal Piers Park 83
Owls Head Park and American Veterans Memorial Pier 83
Calvert Vaux Park 83
Plumb Beach 83
Four Sparrow Marsh 84

3 QUEENS 85

Key Sites 87

Jamaica Bay Wildlife Refuge 87
Big Egg Marsh, aka Broad Channel American Park 93
Jacob Riis Park and Fort Tilden 94
Breezy Point 99
Edgemere Landfill 102
Forest Park 104
Queens Botanical Garden 108
Alley Pond Park and Oakland Lake 109

Other Places to Find Birds in Queens 115

Baisley Pond Park 115
Rockaway Beach Endangered Species Nesting Area 116
Kissena Park and Corridor 116
Flushing Meadows Corona Park 117
Willow Lake 118
World's Fair Marina 118
Highland Park and Ridgewood Reservoir 118
Cemetery of the Evergreens 119

4 THE BRONX 121

Key Sites 121

Pelham Bay Park 121
New York Botanical Garden and Bronx Zoo 127
Van Cortlandt Park 131

Other Places to Find Birds in the Bronx 136

Woodlawn Cemetery 136
Wave Hill 137
Riverdale Park / Raoul Wallenberg Forest Preserve 138
Spuyten Duyvil Shorefront Park 138
North Brother and South Brother Islands 138

5 STATEN ISLAND 139

Key Sites 141

Clove Lakes Park 141
Great Kills Park 145
Blue Heron Park Preserve 149
Wolfe's Pond Park and Acme Pond 151
Lemon Creek Park 154
Mount Loretto Unique Area and
    North Mount Loretto State Forest 156
Long Pond Park 159
Conference House Park 160
Clay Pit Ponds State Park Preserve 162
High Rock Park and Conservation Center and Moses Mountain 164
Mariner's Marsh Park 166
Goethals Pond Complex, Including Bridge Creek,
    Old Place Creek Park, and Goethals Pond 168
Snug Harbor and Allison Pond Park 170
Willowbrook Park 171
Miller Field, Midland Beach, and South Beach 172

Other Places to Find Birds on Staten Island 173

King Fisher Park 173
Oakwood Beach 174
Moses Mountain 174
Fort Wadsworth 174
Silver Lake Park 175
Tottenville Train Station 175

Other Greenbelt Parks 176

Reed's Basket Willow Swamp 176
LaTourette Park 176

In Case You Were Wondering 176

Freshkills Park 176
Harbor Herons Complex—Shooters and Prall's Islands 177
Staten Island Ferry 177

6 NASSAU COUNTY 178

Key Sites 180

Jones Beach State Park 180
Point Lookout 185
Nickerson Beach 186
Cow Meadow Park and Preserve 188
Oceanside Marine Nature Study Area 189
Hempstead Lake State Park 192
Massapequa Preserve and Tackapausha Museum and Preserve 195
John F. Kennedy Memorial Wildlife Sanctuary and Tobay Beach 198

South Shore Winter Freshwater Birding 198

Grant Park Pond and Willow Pond 199
Lofts Pond Park 200
Milburn Pond 200
Cow Meadow Park and Preserve 201
Camman's Pond Park 201
Mill Pond Park and Twin Lakes Preserve,
    Including Wantagh Pond and Seaman Pond 202
Massapequa Preserve and Tackapausha Preserve 203

The North Shore 203

Muttontown Preserve 203
Leeds Pond Preserve 206
Sands Point Preserve 207
Whitney Pond Park 207
William Cullen Bryant Preserve 208
Garvies Point Preserve 208
Welwyn Preserve 208
Stehli Beach Preserve and Charles E. Ransom Beach 209
Centre Island Town Park 209
Bailey Arboretum 209
Shu Swamp (Charles T. Church Nature Sanctuary) 210
Upper and Lower Francis Ponds 210
Planting Fields Arboretum State Historic Park 210
Mill Pond in Oyster Bay 210

Sagamore Hill 211
St. John's Pond Preserve 211
Uplands Farm Sanctuary 212

7 SUFFOLK COUNTY 213

Western Suffolk 215

Robert Moses State Park 215
Sunken Forest at Sailors Haven, and Watch Hill 219
Smith Point County Park 220
Wertheim National Wildlife Refuge 222
Captree State Park and Gilgo Beach 224
Caumsett State Historic Park 225
Target Rock National Wildlife Refuge 228
Tung Ting Pond and Mill Pond 229
Sunken Meadow State Park 230
Blydenburgh Park 234
David Weld Sanctuary 234
Connetquot River State Park 235
Bayard Cutting Arboretum State Park 237
Heckscher State Park 237

Central Suffolk County, Including the Grasslands 239

Wading River Marsh Preserve 239
Wildwood State Park 239
Hulse Landing Road 239
EPCAL 241
Calverton Ponds Preserve, Preston's Pond, and Swan Pond 242
"The Buffalo Farm" 242
Golden Triangle Sod Farms 243

The South Fork and Shelter Island 243

Shinnecock Bay and Inlet 243
Dune Road 245
Cupsogue Beach County Park 248
Mecox Bay 250
Mashomack Preserve on Shelter Island 251

The North Fork 253

Orient Point County Park 254
Plum Island 254
Orient Beach State Park 254
Ruth Oliva Preserve at Dam Pond 255
Inlet Pond County Park 256
Moore's Woods 256
Arshamomaque Preserve 256
Arshamomaque Pond Preserve 256
Cedar Beach County Park 257
Goldsmith's Inlet Park 257
Nassau Point of Little Hog Neck 257
Downs Farm Preserve 257
Marratooka Lake Park (Marratooka Pond) 258
Husing Pond Preserve 258
Laurel Lake 258

Montauk Peninsula 258

Montauk Point State Park and Camp Hero State Park 258
Shadmoor State Park and Ditch Plains 263
Hither Hills State Park 264
Hook Pond 265

8 SPECIES ACCOUNTS 266
Rarities 291
Accidentals 294

Bibliography 297
Index 301

Acknowledgments

This book would not have been possible without the endless support from friends and the help of so many from the birding community. Author/ naturalist and wildlife photographer Stan Tekiela made the generous recommendation that allowed us to have this amazing experience. Wayne Mones, Seth Wollney, and Cliff Hagen unlocked Staten Island. Pat Aitken went above and beyond. Andrew Baksh, Rich Kelly, Joe Giunta, Eileen Schwinn, Nancy Tognan, and Rob and Deborah Wick shared their love and knowledge of Long Island. Annie Barrie helped enormously with Governors Island and Inwood Hill. Tod Winston introduced us to the glories of Woodlawn Cemetery. Without Eyal Megged, the species accounts simply wouldn't have happened. Todd Pover provided valuable information about beach-nesting birds, and Paul Guris whipped the pelagic section into shape. Debbie Becker, Gabriel Willow, Jack Rothman, Jill Weber, Paul Sweet, and the members of the Brooklyn Bird Club are generous evangelizers of birding. Ellen Manos and Peter Skoufalos and Stephen and Laure Moutet Manheimer provided luxurious accommodations during the days of Long Island research.

We are also grateful to the many readers and friends who went over the manuscript and made essential suggestions: Bobbie Bristol, Jo Mader, Shellie Karabell, Jeff Rosenheim, and Mary Tannen. Nancy Ward, Ginny Carter, and Jenny Maritz are the best birding buddies anyone could ever have. We relied on the help of many from Audubon New York: Erin Crotty, Mike Burger, Gini Stowe, Margot Ernst, and the ever-patient Jillian Liner. We also relied on colleagues at New York City Audubon: Kathryn Heintz, Susan Elbin, Debra Kriensky, Barbara Lysenko, Darren Klein, Andrew Maas, and Joe O'Sullivan.

Our skillful editors and designers at the University Press of New England could not have been more patient and helpful to two novice writers. Thank you Phyllis Deutsch, Mindy Basinger Hill, Lauren Seidman, Glenn Novak, and Amanda Dupuis for giving us the opportunity to see New York City in a whole new light, and thank you Eric A. Masterson for showing us the way.

Jaya Sahihi and Phoebe Rosenheim helped finish the manuscript. Ted Goodman did the expert indexing.

All photographs in this book are by Deborah Rivel. Map research and reference materials provided by Kellye Rosenheim.

BIRDWATCHING IN NEW YORK CITY
AND ON LONG ISLAND

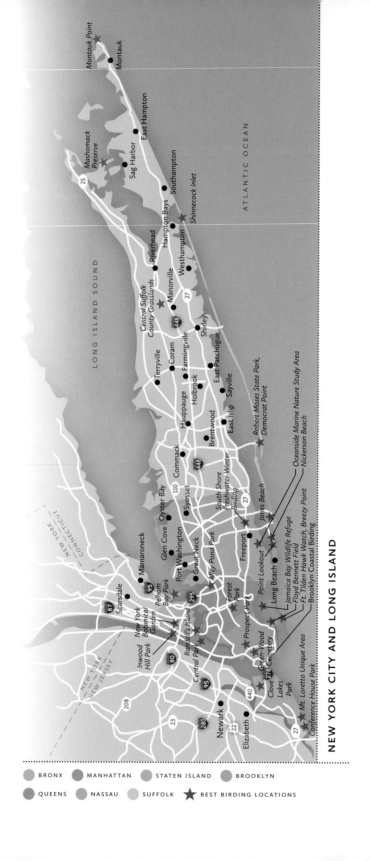

NEW YORK CITY AND LONG ISLAND

Montauk Point
Montauk
East Hampton
Mashomack Preserve
Sag Harbor
Southampton
25
Hampton Bays
Shinnecock Inlet
Riverhead
Westhampton
ATLANTIC OCEAN
Central Suffolk County Grasslands
Manorville
27
LONG ISLAND SOUND
448
Shirley
Terryville
Coram
Farmingville
East Patchogue
Holbrook
Hauppauge
Sayville
Robert Moses State Park, Democrat Point
Brentwood
East Islip
Commack
448
Oceanside Marine Nature Study Area
Nickerson Beach
110
South Shore Freshwater Winter Birding
Glen Cove
Oyster Bay
Syosset
27
Jones Beach
Mamaroneck
Port Washington
Great Neck
Valley Pond Park
Freeport
Point Lookout
Long Beach
Jamaica Bay Wildlife Refuge
Floyd Bennett Field
Ft. Tilden Hawk Watch, Breezy Point
Brooklyn Coastal Birding
NEW YORK
CONNECTICUT
Scarsdale
95
Pelham Bay Park
295
Forest Park
Prospect Park
87
New York Botanical Garden
Inwood Hill Park
Randall's Island
Central Park
Green-Wood Cemetery
80
278
NEW YORK
NEW JERSEY
208
Clove Lakes Park
Mt. Loretto Unique Area
440
Conference House Park
23
95
Newark
280
Elizabeth
222
27

NEW YORK CITY AND LONG ISLAND

- ● BRONX
- ● MANHATTAN
- ● STATEN ISLAND
- ● BROOKLYN
- ● QUEENS
- ● NASSAU
- ● SUFFOLK
- ★ BEST BIRDING LOCATIONS

Introduction

We're often asked, "Are there really any birds here?" In Manhattan, it seems at first glance rather unlikely, with the city almost completely paved over and, further afield, the endless suburban sprawl. But, actually, we get a lot of interesting and, at times, rare birds both in New York City's five boroughs and on Long Island. And because there are a lot of people looking for them, we know quite a bit about their favorite haunts.

So, why do birds come here? New Yorkers have the great good fortune to be on the Atlantic Flyway, one of four migration routes in North America that birds use as a highway on their twice-yearly commute to and from their breeding grounds. The Atlantic coastal area, including New York City and Long Island, encompasses rich and diverse habitats. Over five hundred bird species take advantage of this area to rest and bulk up before they fly onward, or to stay and raise a family. It is therefore an ideal area to find birds, and sometimes they are found in staggering numbers. Waterfowl in particular can be seen from Brooklyn and Montauk in rafts numbering in the thousands to tens of thousands. Forty species of migrating warblers can be found in Central Park, Prospect Park, and at Clove Lakes. These are just examples of the richness and variety of birdlife found in the New York City area, and it can be an eye-opening experience to discover the diverse locations in which these birds are found.

This book tries to give an honest, though admittedly subjective, opinion about the relative merits of each location that we cover. It is in no way comprehensive, as there are numerous pocket parks and favorite birding spots that locals have discovered while scouring their neighborhoods for birds. But we did try to cover the most important as well as most interesting locations. Some of them may be a bit unconventional, but we know that birdwatchers range from "listers" who want to add as many birds as possible to their life list, to people for whom the Zen of birding is an experience unto itself, to others who aren't really that interested in birds but have family members or friends who drag them along on outings. We try to accommodate all potential birding needs, but ultimately the criteria for inclusion was whether the place really had birds and whether we would go there if we weren't writing the book.

We feel compelled to state that in researching this book we made discoveries that gradually changed our entire view of the place we call home. As longtime Manhattanites, we have had a fairly egocentric view of New York in probably every way, and certainly as birders. Having easy access to Central Park does that sometimes. Now having had the opportunity to visit many places we had never been to, we have a more expanded view of where we live and a greater appreciation for the numerous areas where birds find refuge and sustenance on their migrations, and where they make a year-round living. It's also apparent that, as everywhere in our world today, some of these places are becoming less hospitable to birds and other wildlife by succumbing to adverse human-induced changes.

We are very grateful to the many helpful birders who shared their knowledge with us. Birding is a treasure hunt. We hope that with this book you enjoy the quest, and that you find your own Edens and your own avian treasures in New York City and on Long Island.

A Birder's Year

JANUARY AND FEBRUARY

While remaining songbirds are often in muted colors, the waterfowl are in their breeding plumage and a counterpoint to an otherwise gray season. By February the numbers of individuals can be at their peak. Migrating hawks start to return. Red-tailed Hawks and Peregrines begin mating in late winter.

MARCH

Migration to and from this area begins. Waterfowl are on the move, and many are gone by the end of the month. Early migrants such as American Woodcock, Eastern Phoebe, Golden-crowned Kinglet, and the first sparrows make their way here. Piping Plovers and American Oystercatchers arrive on beaches and begin staking out their territories.

APRIL

Migration is well under way and getting better as we approach the end of the month. The first warblers appear (Pine, Palm, Yellow-rumped), followed by the rest and other songbirds. Osprey arrive. Shorebird migration is in progress and builds in May.

MAY

From about May 4 to May 15 we're at the peak of songbird migration. If you have the luxury of making last-minute plans, you'll want to choose a day when the overnight winds have been out of the south. Clear skies are also a benefit. Be sure to check the web and social media for rarities. The last of the migrating warblers is the Blackpoll; when the females appear, you know that the end of warbler migration is near. Wood Duck chicks begin jumping out of nests. Red Knots arrive in May as Horseshoe Crabs spawn.

JUNE

Songbird migration is effectively over, and nesting season has begun. It's a great time to get out and see some of the birds you missed earlier. *Please remember that the birds you seek are in the very serious and delicate business of laying eggs and raising young. Be ever mindful of this and respect their need for privacy and non-intrusion.* In June, raptor chicks start fledging, and in late May and early June you will see Mallard ducklings trailing their mothers. Shorebirds that nested in April and May will start hatching chicks late May and June. Look for Piping Plover and American Oystercatcher babies. Least Terns are starting to hatch, and their parents are providing hawklike protection. Black Skimmers begin nesting—often on the same beaches previously used by plovers and oystercatchers.

JULY

Nesting season continues. Parents and babies will be out and about. Black Skimmer chicks are hatching and on the beach through August. Shorebirds are on the move again, and the first migrating songbirds, generally warblers nesting in the region, are beginning to make their way back. We know this because local species that have been absent from urban parks during the breeding season, like Yellow Warbler and American Redstart, reappear in Central Park.

AUGUST

Shorebird migration comes to its peak August through September. Songbird migration builds. Caspian and Royal Terns start gathering with their young in preparation for migration. Swallows assemble and commence migration.

SEPTEMBER

Songbird migration peaks as birds are now heading south in large numbers, as are raptors. In general, the first birds through in the spring—sparrows and early thrushes—are the last birds through in the fall. Ospreys leave before the chill.

OCTOBER

Songbird migration noticeably tapers off. Raptor migration is in full swing (see our Fall Hawk Watches section). Shorebirds like Sanderlings are seen in large numbers on beaches now before migrating. Waterfowl begin their return.

NOVEMBER

Fall sparrows are at their peak. Raptor migration tapers off, but continue to check beaches for migrating Peregrines early in the month. By the end of November, overwintering birds are essentially all that are left. Dark-eyed Juncos and White-throated Sparrows spend the winter here and leave in spring. Waterfowl season gains momentum. Seabird migration continues with birds such as scoters and gannets.

DECEMBER

Birders flock to feeders and find consolation in the Christmas bird counts. "Gullheads" are scanning tirelessly for rarities, so if you've never seen an Iceland or Glaucous Gull, check reports. Occasionally an irruptive species like crossbills, redpolls, or Snowy Owls overwinters here—it all depends on the scarcity or abundance of their normal food source. Generally, freshwater lakes and ponds do not stay open all winter, and birds that use this habitat move south or to saltwater. However, opportunities for seeing waterfowl are abundant. This is the best time to look for owls. The online reporting of their roosts is frowned upon, as these birds are easily disturbed, so the best way to find out about them is by word of mouth.

The Best

If you have only a limited amount of time to see birds in the New York City area, or are planning a trip and want to make birding part of the itinerary, we thought it might help to highlight the places you are not going to want to miss and the seasons in which they are at their best (see next page). Many of these places are obvious, as they are famous for birds in a particular borough or county. We took the liberty of adding a few subjective choices—places that have interesting and/or a large number of birds but are somewhat off the radar.

New York Harbor Boat Tours

There are many ways to see New York Harbor by water, but you may want to consider taking a tour especially designed to see birds and wildlife. These tours last a few hours and commonly cruise past the East River's North and South Brother Islands, lower New York Harbor's Swinburne and Hoffman Islands, or Jamaica Bay. In summer, you'll see nesting waterbirds, the city at its most scenic, and often spectacular sunsets. In winter, you'll get interesting waterfowl and Harbor Seals. These are great family outings and a wonderful way to see New York City from a different perspective. Tourists find them fascinating, but don't let them have all the fun. This is a must-do.

Pelagic Birding

While New York City might not be known for encounters with birds that stay offshore their entire lives, only coming on land to breed, sighting pelagic birds is easy if you take a boat trip devoted to that purpose. Several excursions run annually, generally departing from a port on the south shore of Long Island and heading toward the waters over the Hudson Canyon, a 135-mile-long submarine trench beneath the Atlantic Ocean. This underwater geography is ideal habitat for the marine creatures that attract seabirds. Depending upon the season, these trips try to locate birds feeding along this and other submarine structures, around commercial fishing boats, and beyond the edge of the continental shelf that runs roughly parallel to Long Island.

As on land, seabirding has its seasons, with species like Cory's, Great, and Audubon's Shearwaters appearing in summer, and alcids like Razorbill,

Best of New York City and Long Island Birding

County	Location	Spring	Summer	Fall	Winter
MANHATTAN	Central Park	•		•	
	Inwood Hill Park	•	•	•	
	Randall's Island wetlands	•			
BROOKLYN	Prospect Park	•		•	
	Green-Wood Cemetery	•		•	
	Floyd Bennett Field				•
	Coastal Birding				•
QUEENS	Jamaica Bay Wildlife Refuge	•	•	•	
	Ft. Tilden Hawk Watch			•	
	Breezy Point	•	•	•	•
	Forest Park	•		•	
	Alley Pond Park	•		•	
BRONX	Pelham Bay Park	•	•		•
	New York Botanical Garden	•		•	
STATEN ISLAND	Clove Lakes Park	•		•	
	Mt. Loretto Unique Area	•		•	•
	Conference House Park	•		•	•
NASSAU	Jones Beach	•	•	•	•
	Point Lookout	•		•	•
	Nickerson Beach		•		
	Oceanside Marine Nature Study Area		•		
	South Shore Freshwater Winter Birding				•
SUFFOLK	Robert Moses State Park, Democrat Point	•	•	•	•
	Central Suffolk County Grasslands			•	•
	Shinnecock Inlet				•
	Mashomack Preserve	•	•	•	
	Montauk Point				•
NEW YORK HARBOR BOAT TOUR			•		•

Common Murre, and Atlantic Puffin in the winter. The thrill of seabirding is that you never know what you'll find. In just the past ten to fifteen years, trip organizers have confirmed New York's first Fea's Petrel and Western Gull records, determined that Leach's and Band-rumped Storm-Petrels are regular summer visitors beyond the edge of the continental shelf in water over a thousand fathoms (six thousand feet), set state record high counts for several species, and discovered that the location of Dovekies in the winter can usually be predicted by water temperature.

If you do think about joining one of these voyages, be prepared for the basic conditions provided by a fishing boat and an often many-hours-long trip, sometimes overnight, on open water. During the warmer months this is necessary to get far enough out to sea to gain views of desired species, although in winter some can be seen on shorter trips. If you're game, it's a great opportunity to add birds to your life list, as the range of birds seen cannot be duplicated onshore. Notices of these trips are often posted online through Listservs like NYSBirds-L and via sponsoring bird clubs and environmental organizations.

Fall Hawk Watches

Technically speaking, the best hawk watches are not in southeastern New York but close by in New Jersey. Palisades Park has two: the State Line Hawk Watch in Alpine and Hook Mountain in Upper Nyack. There is also an excellent one at Sandy Hook in the National Gateway Recreational Area. They're not far by car, although finding the right spot at Hook Mountain is pretty tricky. It's best to go first with someone who knows where to park and how to find the path through the woods.

In the area covered by this book, our favorite is the watch atop Battery Harris East in Fort Tilden. This large decommissioned gun emplacement rises above the dunes and gives spectacular 360-degree views for miles. (See our section on Fort Tilden in the Queens chapter.)

Other places where you may find migrating raptors but not in large numbers are Belvedere Castle in Central Park, Manhattan; at Wave Hill and Pelham Bay Park in the Bronx; Moses Mountain in High Rock Park, Silver Lake Park, and at Spring Pond in Blue Heron Park on Staten Island; and at Robert Moses State Park near the lighthouse and at Democrat Point in Suffolk County.

For best results, pick a clear day with winds from the north or northwest. For a list of the best hawk watches in the Northeast, go to battaly.com.

Peregrine Falcons in New York City

New York City has a special role in the early reintroduction of Peregrine Falcons, which are now found again across North America. Between egg collectors, hunting, and DDT, which thinned their eggshells, Peregrines were in a lot of trouble in the 1950s and 1960s. Until they were protected by the Endangered Species Act, Peregrine Falcons were in a near-fatal downward spiral.

In the 1970s, falconer Heinz Meng and scientist Tom Cade from the Cornell Hawk Barn began raising captive-bred Peregrine chicks and releasing them. Great Horned Owls killed the young birds, so Meng and Cade looked for a predator-free urban environment that mimicked the birds' natural habitat. The Con Ed building in Manhattan was chosen as the hacking site, where juvenile birds were given food until they learned the artful skill of hunting and were able to survive on their own. Wide city avenues imitated their natural canyons, and the geography provided open space for hunting. Most importantly, there were no predators.

The concept was so successful that now there are dozens of Peregrines nesting around the city. The New York Department of Environmental Protection sets up and monitors nesting boxes, and the birds are responding, building nests on every bridge into Manhattan, on church spires, hospitals, even on Wall Street skyscrapers.

The Bird's Point of View

For a bird, flying over any built-up environment has its challenges. Add the marathon-like stress of migration, and it can be fatal. Reflective glass lures birds to their death by collision. When migrating at night, songbirds often get disoriented by the lights in tall buildings and exhaust themselves in confused flight, often striking windows or using up all their energy until they sink to the ground from exhaustion. Once there, they are vulnerable to predators and injury. City lights also make it possible for daytime predators to hunt migrating birds at night, removing the natural protection of darkness. And some birds, having flown all night, can't find a suitable habitat and simply have to put down when they run out of gas. In New York City, woodcocks seem to have more problems than most other birds, and Bryant Park catches a number of them and gives them some refuge despite the crowds. More likely, you'll find songbirds like hummingbirds, kinglets, or juncos stunned or exhausted on the sidewalk.

Fortunately, a law passed in New York State prohibits state-run buildings from leaving the lights on at night when no one is occupying the office. In New York City, more and more tenants are agreeing to a "Lights Out" program as well. This benefits both migrating birds and energy conservation.

Even with this accommodation in place, a migrating bird still needs a safe place to fuel up. And given how few opportunities exist in the city, when birds finally do come upon a park in the morning, they may be found in large numbers and feeding out in the open.

Injured or Dead Birds

If you find an injured bird, call and arrange to take him to the Wild Bird Fund, www.wildbirdfund.org, 565 Columbus Avenue (at 87th Street), New York, New York 10024, 646-306-2862. They and the New York City Audubon Society will also know whom to call if you're in one of the outer boroughs or on Long Island (212-691-7483). If you find a dead bird, the conservation scientists at NYC Audubon want to know about it. Use their D-bird reporting app to add this critical information to their database, www.d-bird.org. They can also help if you have questions about what to do about a bird in distress, 212-691-7483.

Important Bird Areas

You will see references to Important Bird Areas (IBAS) throughout this book. The purpose of the IBA program is to scientifically identify places that are especially critical because they support large congregations of birds, or birds that are of conservation concern. Many birds at those sites are vulnerable to habitat loss, disturbances, and other threats, and the IBA program strives to reduce those threats and ensure the protection of those places in the future. The IBA program was begun by the National Audubon Society and is international in scope. The IBAS in this book have been identified by the scientists at Audubon New York. It should be noted that while the IBAS have been identified, and conservation practices are in place at most of them, the number and diversity of species found at any particular one may vary over time as a result of changes brought by nature or human intervention.

Respect for Nesting Shorebirds

As birdwatchers we all want to protect the birds we seek. There are commonsense ethics that apply to watching birds in general (see the ABA Code of Ethics below), but having information on the particular needs of nesting shorebirds might be helpful when you visit. In general, beach-nesting birds have a rough time. We all share the nice flat sand (they nest and raise their chicks on the same beaches where we play volleyball, run our dogs, and ride ATVs) and shallow tidal pools (they feed their chicks in them, we play in them). So it's not a great surprise that some of these birds are becoming increasingly rare, and others, like Piping Plover, are endangered. Here are a few things to take into consideration:

Roped-off areas outline the nesting areas where the eggs are laid on the sand in a slight depression called a "scrape" and are meant to keep them free from humans and dogs. The nest is nearly imperceptible, and once hatched, the chicks blend into the sand so well that a baby Piping Plover that isn't moving can be barely visible. But Piping Plovers are precocial and will start running around almost immediately after hatching. They also go out of the roped-off area to the water's edge to feed with their parents, who battle ghost crabs hunting their chicks, and people, dogs, gulls, crows, foxes, and cats—you get the picture. The parent plovers work together to keep their energetic chicks herded and guarded, aggressively defending them and using a variety of means. If you hear an intense and constant peeping or see a Piping Plover that appears to have a broken wing luring you away, it means a chick is really close and probably very still. Move away, but be alert so that you don't step on one hiding in a depression or tire track.

Unlike plovers, Least Terns are colonial nesters, and often these colonies are huge and span several beaches, with hundreds of pairs of birds. The roped-off areas may be divided by paths to the beach through the colony. These birds are aggressive and target accurately from above with a purple substance that doesn't wash out easily. Give yourself and them a break and move quickly out of range.

Piping Plovers defending chicks from American Oystercatcher

Many beach-nesters do so on prime real estate. When changes are made to their breeding grounds such as needed beach replenishments and dune reconstruction after a hurricane, neither the needs of humans nor birds are always part of the program. The results can be devastating, especially to the birds. Their nesting habitats and prey base are often destroyed for many years, and it can take a toll on the species as a whole. As a birder, it also changes where you can find the birds and how many of them there are. If work is under way or has been done on any of the beaches that you plan to visit, check to see the current populations and locations of birds first.

As with all birds, good views and especially photos of them take time, patience, and respect for the birds' needs. If you give them space and the chance to feel comfortable, you will be amply rewarded.

ABA Birding Ethics

The American Birding Association (www.aba.org) has a list of birding ethics that we subscribe to and recommend to our readers to make a pleasanter experience for you, your fellow birders, and—most important of all—the birds you are watching. They are as follows:

Everyone who enjoys birds and birding must always respect wildlife, its environment, and the rights of others. In any conflict of interest between birds and birders, the welfare of the birds and their environment comes first.

1. Promote the welfare of birds and their environment.

   a. Support the protection of important bird habitat.

   b. To avoid stressing birds or exposing them to danger, exercise restraint and caution during observation, photography, sound recording, or filming. Limit the use of recordings and other methods of attracting birds, and never use such methods in heavily birded areas, or for attracting any species that is Threatened, Endangered, or of Special Concern, or is rare in your local area. Keep well back from nests and nesting colonies, roosts, display areas, and important feeding sites. In such sensitive areas, if there is a need for extended observation, photography, filming, or recording, try to use a blind or hide, and take advantage of natural cover. Use artificial light sparingly for filming or photography, especially for close-ups.

   c. Before advertising the presence of a rare bird, evaluate the potential for disturbance to the bird, its surroundings, and other people in the area, and proceed only if access can be controlled, disturbance minimized, and permission has been obtained from private landowners. The sites of rare nesting birds should be divulged only to the proper conservation authorities.

   d. Stay on roads, trails, and paths where they exist; otherwise keep habitat disturbance to a minimum.

2. Respect the law, and the rights of others.

   a. Do not enter private property without the owner's explicit permission.

   b. Follow all laws, rules, and regulations governing use of roads and public areas, both at home and abroad.

   c. Practice common courtesy in contacts with other people. Your exemplary behavior will generate goodwill with birders and non-birders alike.

3. Ensure that feeders, nest structures, and other artificial bird environments are safe.

   a. Keep dispensers, water, and food clean, and free of decay or disease. It is important to feed birds continually during harsh weather.

   b. Maintain and clean nest structures regularly.

   c. If you are attracting birds to an area, ensure the birds are not exposed to predation from cats and other domestic animals, or dangers posed by artificial hazards.

4. Group birding, whether organized or impromptu, requires special care. Each individual in the group, in addition to the obligations spelled out in Items #1 and #2, has responsibilities as a Group Member.

   a. Respect the interests, rights, and skills of fellow birders, as well as people participating in other legitimate outdoor activities. Freely share your knowledge and experience, except where code 1(c) applies. Be especially helpful to beginning birders.

   b. If you witness unethical birding behavior, assess the situation, and intervene if you think it prudent. When interceding, inform the person(s) of the inappropriate action, and attempt, within reason, to have it stopped. If the behavior continues, document it, and notify appropriate individuals or organizations.

5. Group Leader Responsibilities

   c. Be an exemplary ethical role model for the group. Teach through word and example.

   d. Keep groups to a size that limits impact on the environment and does not interfere with others using the same area.

   e. Ensure everyone in the group knows of and practices this code.

   f. Learn and inform the group of any special circumstances applicable to the areas being visited (e.g., no tape recorders allowed).

   g. Acknowledge that professional tour companies bear a special responsibility to place the welfare of birds and the benefits of public knowledge ahead of the company's commercial interests. Ideally, leaders should keep track of tour sightings, document unusual occurrences, and submit records to appropriate organizations.

Please follow this code and distribute and teach it to others.

Safety

It doesn't seem right somehow that you should be concerned about your safety when you are enjoying nature and following birds. Sadly these days, no matter where you are birdwatching, common sense and alertness are important. The areas covered in this book range from highly urban neighborhoods to remote, wild preserves. In some you might not feel (or be) safe if you are exploring them alone.

That being said, if you have visions of New York City as a hotbed of crime, your information is outdated. New York City's crime rate has been dropping for over twenty years, and given the millions of people who live here, it is a relatively safe place. Regardless of where you bird, our advice remains the same: be alert no matter where you are, and we always recommend birding with a friend. Period. This is probably the kind of advice your mother would have given you, but, unlike your mother, we fully acknowledge that this is not always possible. We also know that there are times that birding alone will be the best part of your day and a restorative endeavor.

You may see a few somewhat dicey spots described in this book. We considered not including them, but since they are known birding locations, we put them in—each with a warning about safety. In some instances, only particular sections of a park are less safe. But if you discover yourself in a place that makes you uncomfortable, use common sense and just leave.

A couple of other things to be alert for are poison ivy—it's everywhere!—and ticks, which are abundantly found in grassy and forested areas on Long Island, Staten Island, and in a few of the wilder parks in the other boroughs. Learn to identify poison ivy's three-leafed configuration and avoid it. Check yourself for ticks upon coming home. The best remedy is prevention, so always wear long sleeves, and tuck your pants into your socks. A hat with a large brim will help keep ticks off your head and face. Adhering to these sartorial guidelines may make you feel a little silly but will greatly reduce the possibility of contact with either of these undesirable elements.

Common Conundrums

Ever had that horrible realization, as you're wrapping up an exciting day of birding, that you've missed something important? There are countless opportunities for cases of mistaken identity between similar birds, and we have all experienced this, especially when we're in unfamiliar territory. So that that won't happen to you, here are a few common puzzlers—similar birds whose distinguishing characteristics you might want to review beforehand in your favorite field guide:

Common vs. Forster's vs. Arctic Tern The Arctic Tern is rare and usually
found only in summer.

Double-crested vs. Great Cormorant The Great Cormorant is typically seen

The puzzling yellowlegs. Greater or Lesser? Even the experts can't always agree.

only in coastal areas and pretty much vanishes during the breeding season.

Greater vs. Lesser Scaup

Least vs. Semipalmated vs. Baird's vs. White-rumped vs. Western Sandpiper

Short-billed vs. Long-billed Dowitcher The latter is really only seen in our area in the fall.

Greater vs. Lesser Yellowlegs

Hudsonian vs. Marbled Godwit

Acadian vs. Alder vs. Willow Flycatcher The Willow is most common; the Alder is the least common.

American Crow vs. Fish Crow

Northern Waterthrush vs. Louisiana Waterthrush Keep in mind that the Louisiana migrates into our area in spring generally before the Northern, but that there is overlap, and therefore timing is not definitive.

Blackpoll vs. Bay-breasted Warbler, in fall.

Common vs. Boat-tailed Grackle

Black-crowned vs. Yellow-crowned Night-Heron, juveniles.

Sharp-shinned vs. Cooper's Hawk

Glossy vs. White-faced Ibis The latter is quite rare.

Swainson's vs. Gray-cheeked vs. Bicknell's Thrush Of course, we also commonly have Veery and Wood and Hermit Thrushes. Of the former threesome, however, the Swainson's is the most common. The Bicknell's is so uncommon that no one will believe you without a positive song identification.

Van
Cortlandt
Park

Bronx River Pkwy.

HACKENSACK

RIVERDALE

95

95

Begin
Hudson
River
Greenway
Tour

Inwood
Hill Park

Ft. Tryon
Park

Sherman
Creek and
Swindler
Cove

New York
Botanical Garden

1

FORT LEE

George
Washington
Bridge

1

B R O N X

WASHINGTON HEIGHTS

New Jersey

9A

87

278

95

HUNTS
POINT

"The Drip"

Morningside
Park

Riverside Park

HUDSON RIVER

FDR Drive

278

RFK
Triborough
Bridge

Central
Park

Randall's
Island

LaGuardia
Airport

1

Carl Schurz Park

495

Lincoln Tunnel

West Side Hwy.

WEEHAWKEN

HELL'S
KITCHEN

ASTORIA

278

Peter Detmold Park

QUEENS

HOBOKEN

High Line

Bryant Park

Madison Square Park

Midtown Tunnel

Union Square Park

Holland Tunnel

Washington
Square Park

Prospect Park

GREENPOINT

278

EAST RIVER

The
Battery

Staten
Island Ferry

Water Taxi
Dock

GLENDALE

Jackie Robinson Pkwy.

Governors
Island

Brooklyn-Battery/
Hugh L. Carey Tunnel

B R O O K L Y N

MANHATTAN

1 Manhattan

An island twelve miles long and three miles wide, encrusted with skyscrapers and directly under the Atlantic Flyway: this might be a description of Manhattan from a birdwatcher's perspective. Migrating birds would probably note only the big rectangle of green known as Central Park, the refuge they have been waiting for after a long night of flying.

Even though we record that it was settled by the Lenape and "discovered" by Giovanni da Verrazano, the area now known as Manhattan has operated for millions of years as a stopover in the twice-yearly migration of birds. And yet if you ask most New Yorkers, they are completely unaware there are birds here other than pigeons. They are also happily unaware how important the small parks found on this island are and what a high diversity of species rely on them.

Its original name, Mannahatta, means "island of many hills." It might be difficult to see these hills for the tall buildings, but this densely populated economic powerhouse only a few hundred years ago was a wilderness of forests, streams, and wetlands. One little piece of tamed wilderness that's left is the crown jewel, Central Park. Given the scarcity of resources, birds concentrate on what's available to them, meaning that high densities of birds are found here in small spaces. With millions of birds on the Atlantic Flyway overhead, it's no surprise that Central Park is considered one of the top birding hot spots in the United States. It is a delight to explore, and we can only imagine how enticing it looks to a tired migrant.

Birding Manhattan can all be done by public transportation and can be a satisfying and varied experience. *Central Park* alone has numerous diverse habitats, but there are also other places to visit. *Inwood Hill Park* is known for its mature forest and good shorebird habitat, and has, in fact, the largest wetland in Manhattan, *Muscota Marsh*. You have the option to bike and bird the West Side via the *Hudson River Park Greenway*. Smaller parks, like *Bryant*

Red-tailed Hawk with prey

Park, may catch exhausted migrators or harbor the odd rarity. For something a little different, travel out to *Randall's or Governors Islands*. Randall's has a lovingly restored natural area, and Governors, still a work in progress, has a nesting Common Tern colony in summer.

Wherever you are in Manhattan, keep your eyes and ears open for signs of birds, even in unexpected places. If you are alert, you will discover they are all around you. Listen in the morning and early evening in summer for the chatter of Chimney Swifts overhead; kestrels nest everywhere and can be heard calling; Red-tailed Hawks and Peregrines are year-round residents and are often seen flying or in parks. And all over Manhattan, wherever there is a bit of nature, expect to see and hear backyard birds. The backyards here may be smaller than in most other places, but many birds have adapted. Mourning Doves nest in pots and bushes, Mallards on roof terraces, Red-tails on fire escapes. Cardinals and Blue Jays call along tree-lined streets. Mockingbirds mimic car horns. White-throated Sparrows relearn their song every spring in sheltered private gardens. Egrets, herons, cormorants, and gulls are always overhead.

Birding Manhattan also means meeting other birders—and they are everywhere. You can expect to get advice, whether you need it or not, on where to find the warbler just seen. You can also just follow the crowd to a rarity. In many places, you'll find people birding their way across the parks to work. It's an opportunity to make a much-needed connection with nature, and Manhattan birders are happy to show you exactly how to do it.

Central Park

The highlight of Manhattan birding is Central Park, the incomparably beautiful manufactured wild space designed by Frederick Law Olmsted and Calvert Vaux in 1858. During migration, its 843 acres of forest, freshwater, meadows, and open spaces amid the endless concrete and fragmented plantings are irresistible to birds. Its pleasures and attractions are irresistible to the rest of us as well. The vision Olmsted and Vaux had was simple—to transform the rocky land into a combination of parkland and wilderness where, upon entering, anyone could feel a release from the pressures of city life with a stroll through the wilderness. This respite designed for humans is such a haven for birds that it is considered to be one of the prime migration hot spots in the United States. It is an Important Bird Area, and forty species of warbler have been recorded here over time. On a busy day in May, you can easily expect to see over fourteen warbler species. Visitors flock from around the globe to bird its many distinct areas, including the Ramble, Strawberry Fields, the North Woods, and, in winter, the Reservoir.

The birding experience in Central Park is exceptional for the diversity of birds seen, as well as the dedicated local birding community cataloging and reporting every notable sighting. If you go nowhere else in New York City to find birds, you must visit this surprising refuge during migration.

VIEWING SPOTS

Before you start your foray into Central Park, take a look at the map to familiarize yourself with its various parts. The area south of 70th Street is lovely and is home to the Central Park Zoo and a picturesque lake, but if it's birds you are after, you are going to want to be north of 70th. The Ramble is undoubtedly the most popular birding area—a woodland that runs approximately between 72nd and 79th Streets and between the East and West Drives. Complete with a stream, a pond, rocky outcroppings, and rustic bridges, it engages the visitor so completely that it may be difficult to believe you are in the middle of a bustling metropolis. Such was the intent of the design, and it succeeds for the benefit of birds and birders. As its name implies, the Ramble is full of meandering paths and is an easy place in which to get temporarily lost.

Strawberry Fields is a small area near the Ramble that can be surprisingly birdy during migration. There is one path through it, which can be approached from a number of different directions. Strawberry Fields is on the West Side and extends from 72nd Street north to 74th Street, and whatever your schedule, don't linger so long in any other section that you jeopardize your chance to visit here.

More great birding takes place in the North Woods, a mature urban forest and stream surrounded by lawn and edge habitats. It lies mostly to the west and above 100th Street, and while just as productive as the Ramble, it has a wilder, more feral feel. Many consider it best in fall.

Between the Ramble and the North Woods is the Reservoir, primarily a winter waterfowl haven. It's located in the center of the park between 86th and 96th Streets.

The most reliable place to begin your birding tour of Central Park is *the Ramble*.

Happily, there's no way to be both efficient and thorough touring the Ramble, but in the spring and fall start at *Maintenance Meadow*. Morning light warms up the trees on the western side, bringing the songbirds in to feed. From there, try *Tupelo Meadow*, which is dominated by a huge tupelo tree that positively glows red in fall. Now make your way into the woods through the opening in the trees at the southeast side of Tupelo and move along *the Gill*. Circle *Azalea Pond*, and once you've closed the loop, walk a few paces over to *Evodia Field* to look at the feeders. Head south to the overlook at *the Oven*. From this higher vantage point the birds in the trees are at eye level. Walk south down the spit of land known as *the Point*. It can be hit or miss, but on a good day during migration, a wide variety of warblers can be found on its tree-lined paths, which traverse the wooded area and run along the shoreline. Now go back around the Oven and head to *Willow Rock*. Not only will you find birds feeding in the trees around you, but the water below is quiet and secretive enough to attract those desiring a bath.

If you're short on time, walk back into the woods and along the Gill, and try to find *Laupots Bridge*. Birds are found here feeding in the morning, and later in the afternoon warblers and other songbirds bathe in its sheltered stream.

Continue west, exploring the *Upper Lobe.* At Oak Bridge, look for a wood-chip path on your right that takes you north along the east side of the water. There are often nesting birds along this path in summer, and toward the north end there is a large willow at the edge of a lake which is a warbler magnet. Leaving the woods, stop at the lawn across the sidewalk to the north just in case an Indigo Bunting is feeding there. If you are running out of time, take the West Drive south to Strawberry Fields. (See our description below.)

However, if you're at Willow Rock and you've got all day, bird the trees along the north side of *the Lake* by taking a walk along *the Riviera.* Cross *Bow Bridge* and make a right along the southern part of the Lake to *Wagner's Cove.* In fall, this may be the smarter choice. You'll find a path along the water's edge surrounded by trees and bushes that are attractive to birds.

But let's get back to the Upper Lobe described above, where we were before our alternate diversion. From the Upper Lobe, find the West Drive and walk to the right (north and uptown) about two blocks. You'll see the Swedish Cottage, where you'll make a right again to walk past the cottage and enter *Shakespeare's Garden.* In some years, the yew trees have overwintering Screech Owls, so in season, look for telltale droppings on the limbs and ground. Make your way up through the garden and over to *Belvedere Castle.* It has a large lookout over *Turtle Pond* and the island. This is a popular spot with tourists, and the trees to the west of the overlook can have migrants flitting around. They will be at eye level and relatively easy to see. To get well-lit photos of them, you will need to be there in the morning.

CENTRAL PARK

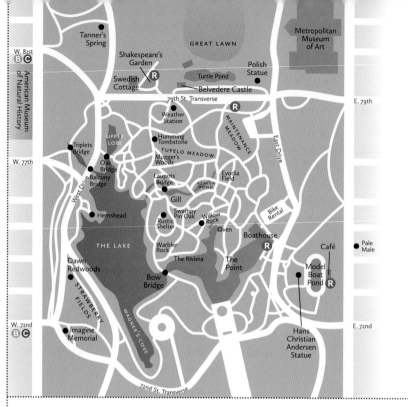

THE RAMBLE

From Belvedere, continue east along the southern shore of Turtle Pond, which can be super-productive, especially in spring. At the end of your shore walk, you'll be near the *Polish Statue*, an equestrian statue of King Jagiello and a birdwatching landmark. The trees around this eastern side of Turtle Pond often have exciting migrants.

This brings you almost full circle back to the Maintenance Meadow, but don't resist the temptation to explore other parts of Central Park. Regardless of your wanderings, make sure you set aside enough time to visit Strawberry Fields.

It may be small, but *Strawberry Fields* can have an intensity that makes it an ideal place to stop by if you don't have a lot of time—but don't blame us if you give in and stay to watch the action on a really good day. There are several ways to get here, and you should take a look at the map for its exact location.

If you are coming from the Ramble, you are approaching Strawberry Fields from the north, passing the vista out onto the Lake at the gazebo, and checking for birds at the water's edge. On your right, you'll see that the West Drive makes a fork at about 74th Street. In the crotch of the fork and a little to the right, take the opening in the wooden fence and climb the hill through the wild rose patch. Once you've reached the summit, pause and face the stand of trees and shrubs to the west, keeping the *Dawn Redwood trees* (the attractive cypress-like ones) on your right. *This is one of the hottest of the hot spots during migration.* In fact, any of the trees that surround you in this open space could have and most likely do have something interesting. Regain the wood-chip path and look for jewelweed. It's where hummingbirds come to feed during fall migration. Keep along this little path until it ends just east of the Imagine memorial to John Lennon. From here you can exit the park easily on 72nd Street.

If you are coming from the south or west, or you only have a few minutes to bird in New York City, pop into Strawberry Fields using the south-end entrance to the same unmarked woodchip path. It's found just east of the Imagine memorial and before you get to the West Drive. The trees and shrubs around here are a favorite spot of ours, and anything from a lurking Ovenbird to Baltimore Orioles can be seen here. Come out onto the lawn, and before the path turns downhill, stop and look to the west. You will face the Dawn Redwood hot spot described in the paragraph above. Despite the crowds of tourists, you should consider taking a friend with you to the woods around Strawberry Fields.

Another good spring migration stop nearby is the top of the *hill behind Falconer's Statue.* Cross the 72nd Street transverse and use the path on the right for an easy walk up. The trees are low and can be productive. Make sure to also check out the lawn on the eastern side.

Ovenbird in Strawberry Fields

To explore the *North Woods*, take a friend and enter at Central Park West and 100th Street. Check *the Pool*, especially the trees on the western side. If it's early in the morning and autumn, continue north to *the Great Hill* for sparrows on the lawn, particularly at the trees' edge to the south and east. Otherwise, make your way along the stream coming out of the Pool, known as *the Loch*, into the woods. The path makes a left-hand turn through *the Ravine*, continues north, and shortly you arrive at the service area behind Lasker Rink and Pool. You can double back under the overpass and take a wooded path back west, but if you have time, continue north, keeping Lasker on your right. Cross the East Drive toward the west, and bird the forest around *the Block House*. Note on the map a little spot called *the Lily Pool*. A carved rock on the footpath near the Block House alerts you to an almost hidden trail into a real urban oasis. It's a great sit spot, especially at the end of the morning when birds are now bathing, drinking, and preening.

The *Reservoir*, late fall through winter and into early spring, has easy-to-spot waterfowl and gulls. Take a walk around the perimeter to see the Hooded Mergansers and Northern Shovelers that congregate on the south side and near the south pump house. The east side of the Reservoir may have Pied-billed Grebe; gulls mostly hang out in the middle. The north pump house also shelters a variety of waterfowl, and look along the western side for Red-breasted Mergansers, Bufflehead, and Gadwall close to the walkway. Be aware that the path around the Reservoir is for runners, so you should walk in the direction of traffic and give them right of passage. On its southern border, on the western side, is a *stand of old oak trees* that exude sap in spring that Blackburnian Warblers really like. The best view is from the wood-and-iron bridge that crosses the bridle path below.

OTHER PLACES TO FIND BIRDS IN CENTRAL PARK

Around the park entrance at Central Park West and 81st Street are a few places to keep on your itinerary. *Triplets Bridge*, north of 77th Street and west of West Drive, crosses the outgoing waters from the Lake. Look for stray migrants at this wooded stream. Just north of the 81st Street playground is a steep hill known as *Summit Rock*. At its base, near the West Drive, is a tiny puddle of water, *Tanner's Spring*, that attracts thirsty birds at all times of the year. It's a good place for birders to stop and rest, too. Photos here can be fairly easy, but the light is relatively dim. On the other side of West Drive is a low outcropping of schist appropriately called *Sparrow Rock*. Visit it early

CENTRAL PARK NORTH WOODS

in spring and later in fall. A bit farther east, stop in at the *Pinetum*, especially the eastern side in fall and winter. It can be productive for sparrows and has been known to have owls in winter, along with Yellow-rumped Warblers and Red-breasted Nuthatches. From the Pinetum, travel south through a wooded area that provides a western border to the ball fields, *Locust Grove and the open lawn to its west*. Sometimes a Red-headed Woodpecker will try to make it through the winter here. It's also a good fall birding spot.

Hallett Nature Sanctuary, located on the East Side between 60th and 62nd Streets just south of Wollman Rink, is one of three forests in Central Park. Originally called "the Promontory" by Olmsted, it was closed to the public in 1934 to provide a bird sanctuary. It reopened in 2013 on a limited basis after an intensive restoration using native plants. It is open to visitors from time to time—check the Central Park Conservancy's website. Hallett provides the backdrop to *the Pond*, which can have interesting ducks in fall and winter, but you're better off looking for those on the Reservoir.

The *Conservatory Garden*, beautiful in spring and summer, also has over-wintering birds, as do the *compost heaps* on the bluff to the west.

Central Park also affords the opportunity to gawk at celebrities. Without a doubt New York City's most famous bird, Pale Male, nests on prime Fifth Avenue real estate with commanding views of the park. You can watch the goings-on at this Red-tailed Hawk nest, which sits on an upper cornice of 927 Fifth Avenue—a residential building on the southeast corner of Fifth Avenue and 74th Street. Go into Central Park to the Model Boat Pond and look for scopes set up on the west side of the pond. Friendly fans of Pale Male are more than happy to let you get a close-up look. You can also find the nest on your own by lining yourself up with the café and putting the Hans Christian Anderson statue behind you. Three limestone apartment buildings rise behind the café. The middle one is a little shorter than the other two. On its top floor are three windows, and the nest is on the curved decorative arch over the middle window. Pale Male and his mate are often found perched nearby.

KEY SPECIES BY SEASON

Spring The season gets under way with the appearance of the first American Woodcock and Eastern Phoebes. Other flycatchers that visit are Eastern Wood-Pewee, Great Crested, and, less commonly, Olive-sided, Yellow-bellied, Acadian, Alder, Least, and Willow (although the Willow Flycatcher is a somewhat common nester just outside the city). Warbling Vireos are common and stay the summer; Blue-headed and Red-eyed Vireos are common migrants; less common are Yellow-throated and White-eyed; and Philadelphia Vireos are quite uncommon. Tree, Barn, and Northern Rough-winged Swallows are among the early migrants and continue through the summer; Blue-Gray Gnatcatchers only migrate through Central Park. Although Golden-crowned and Ruby-crowned Kinglets overwinter here generally, they are rarely seen in the park outside of migration. Thrushes rush through—a better time to enjoy them is in the fall. Indigo Buntings, Orchard Orioles, Scarlet Tanagers and, less often, Summer Tanagers visit—and spring is the only time to see them. Particularly known for the wide variety of warblers seen during migration, Central Park lives up to its reputation as a warbler magnet.

Summer Wading birds continue and include Great Blue Heron and Black-crowned Night-Heron. Chimney Swifts, Eastern Kingbird, and Gray Catbird are present, as are a number of other nesting songbirds, including Baltimore Oriole and Warbling Vireo. Hummingbirds migrate through.

Fall Most of the spring migrators are now returning, and it's a great time for sparrows and more relaxed views of thrushes. Central Park is not the best place to look for migratory hawks, although a watch exists at Belvedere Castle. Red-tails are here year-round, and the more common migrants are Cooper's and Sharp-shinned. Eastern Bluebird is uncommon in Central Park. You have an excellent chance of seeing Hermit, Gray-cheeked, Swainson's, and Wood Thrushes. Brown Thrashers arrive and stay the winter, as do Rusty Blackbird, American Goldfinch, Eastern Towhee, Chipping Sparrow, and several other sparrows. White-throated Sparrows are common in spring, fall, and winter; White-crowned Sparrows migrate through. Purple Finch will make occasional appearances starting in fall and throughout the winter.

Winter The Reservoir is best for waterfowl, although you may see some nice ducks at the Lake and the Pond. Bufflehead, Hooded Merganser, Northern Shoveler, Gadwall, Pied-billed Grebe, and Ruddy Ducks are among the fairly common, with occasional visits from Ring-necked Ducks, Northern Pintails, and Canvasbacks. Eastern Screech and the occasional Great Horned Owl are heard and sometimes seen. An Iceland Gull would be uncommon. In irruption years, there is the possibility for Crossbills and Redpolls. Late winter brings American Tree Sparrow. Black-capped Chickadee, Tufted Titmouse, White-breasted and Red-breasted Nuthatch, Downy Woodpecker, Northern Flicker, Brown Creeper, Carolina Wren, Northern Mockingbird, Cedar Waxwing, House Finch, Northern Cardinal, and Blue Jays are among those present year-round. Red-headed Woodpecker are sometimes seen.

HABITAT

Urban forest and made-made streams, reservoir, ponds, open lawn.

BEST TIME TO GO

In this order: spring, fall, winter, summer.

HOW TO GET TO CENTRAL PARK

By Subway Central Park is literally central to almost all public transportation in Manhattan. To enter from the west, the **B** and **C** make stops along Central Park West, but it's also a short walk from the **1** **2** and **3**. From the

East Side, use the local ⑥ or express ④ and ⑤. Less convenient to the best birding are the Ⓝ Ⓠ and Ⓡ.

By Bus On the West Side, take the M10. To access the East Side and north take the M2, M3, or M4. Crosstown buses are the M66, M72, M79, M86, M96, and M106. Plus there are also plenty of buses from the outer boroughs that run up Madison and down Fifth on the East Side and use Columbus and Broadway on the West Side.

By Car If you drive and need to park, you'll probably find it most convenient to do so at or near the American Museum of Natural History, 81st Street between Columbus and Central Park West.

INFORMATION

http://www.centralparknyc.org
http://www.nycgovparks.org
http://www.philjeffrey.net
https://twitter.com/birdcentralpark

Truly rare birds are reported on the text alert system NYNYBirds. Sign up at https://nynybird.wordpress.com.

Places to eat are dotted around the park. Birders (and dog walkers) converge at the Boathouse Café, which has both an upscale dining room overlooking the Lake and a more utilitarian snack bar with indoor and outdoor tables. Other pleasant spots are the café on the north side of Sheep Meadow and, in the evening, for a drink, the café by the Model Boat Pond. All the above have restrooms for the public. Other comfort stations are located between Delacorte Theater and Shakespeare's Garden, on the north side of Maintenance Meadow, on the Great Hill, in the Dana Center at the Meer, in the underpass by Bethesda Fountain, and at the Conservatory Garden.

Food carts operate throughout the park year-round. Drinking fountains are turned on in spring and off in fall.

Avoid visiting the same day as large outdoor concerts, as sections of the park are cordoned off, and getting around can be frustrating.

Getting lost Olmsted and Vaux designed the Ramble and the North Woods to provide walks in the woods where you lose yourself in nature. While it's easy to get your directions confused, remember that the Ramble is only about seven by three city blocks, and the North Woods is not much larger. Even if you do get lost, once you can see a tall building, head for it and you will eventually wind up on either Fifth Avenue or Central Park West.

Insider tip All lampposts in Central Park carry a four-digit number. The old ones have stamped metal bands; the new ones have white paint. The first two digits correspond to the cross street with which the lamppost is level, so a lamppost with the number 7901 is roughly at 79th Street. Traffic lights do not participate in this program.

Safety Use common sense when talking to strangers, as many different kinds of people frequent the park. We've always felt free to ask other birders questions, not just for directions, but what they've seen today. Even if your interviewee is somewhat reticent, few can resist the urge to brag about a juicy sighting, and having done so are obligated to give directions to it. But even though Central Park is widely visited, some areas in the North Woods and the Ramble, like Warbler Rock, are not always safe if you are alone. Be smart and don't bird here on your own.

Throughout Central Park, there are numerous arcane names for hot spots. Don't be put off because you're not sure where the Swampy Pin Oak is, or some other location you've heard about. Even the most experienced of us lose our grip on the nomenclature now and again.

OTHER THAN BIRDING

Just about every outdoor activity is being pursued here, but some of the more popular things to do are to rent a bike or row a boat. If you are a runner, take advantage of one of the many road races (the New York City Marathon ends on the West Side at 66th Street).

Scope Leave it at home. You'd only take it to look at the waterfowl and gulls on the Reservoir, and that's asking for an accident involving runners on the narrow path.

Photos Lots of opportunities, especially at the Reservoir, Tanner Spring, Maintenance Meadow, and in the Ramble at the Oven, the feeders, the Point, and Laupots Bridge. Some of these locations can have low light.

Inwood Hill Park

Inwood Hill is a beautiful park that offers interesting birding as well the chance to learn about the history of New York City. Situated at the northern

BRONX

Henry Hudson Bridge

Railroad Swing Bridge

SPUYTEN DUYVIL CREEK

MUDFLATS

Flag Pole

Main Entrance

W. 218th St.

Boat Basin

Urban Ecology Center

Indian Road

MUSCOTA MARSH

Spuyten Duyvil Rd.

215th St.

W. 215th St.

BEACH

Great Wall Path

Indian Road Playground

Isham Soccer Field

Isham Ballfields

Lower Spring Rd.

Upper Spring Rd.

MIKE'S MEADOW

Shorakapok Rock

Spring

Dog Run

Veterans Flag Pole

Old Garden

Big Beech Rd.

Middens

Isham St.

Fire Pit

VIEW

Indian Rock Shelters

Emerson Playground

Quarry

OVERLOOK MEADOW

West Ridge Rd.

WEST RIDGE

W. 207th St.

Hudson Soccer Fields

OVERLOOK

North Underpass

A
INWOOD-207TH ST

Hockey Rink

Overpass

W. 204th St.

Glacial Potholes

South Overpass

Payson Ave.

Cooper St.

HUDSON RIVER

Dyckman Fields

Henry Hudson Pkwy. Southbound

Henry Hudson Pkwy. Northbound

Whale Back

Entrance

Academy St.

Entrance

Beak St.

Seaman Ave.

Cumming St.

Broadway

Entrance

Canoe Launch-Dyckman Boat Marina

Dyckman St.

Payson Playground

Ft. Tryon Park

Staff St.

Henshaw St.

A
DYCKMAN STREET

Cloisters ▼

INWOOD HILL PARK

tip of Manhattan, its heavily forested 196 acres of hilly terrain and ridges are a product of ancient glaciers, making for both a surprisingly wild area on the water and a truly urban experience with gorgeous views of the Palisades. After you search for migrating songbirds in the forest and for shorebirds along the mudflats, you can visit the exact location where Native Americans sold Mannahatta to Peter Minuit in 1626. Bald Eagles were released here in a reintroduction program started in 2002, which may account for why you can occasionally see them soaring in the area. The natural salt marshes attract shorebirds, including rare ones like Pectoral Sandpiper, which find the mudflats a gooey smorgasbord during migration. Spring and fall also find a variety of songbirds hunting insects in the trees. Because the forest is so productive, it holds on to migrating songbirds in the spring for about one week longer than Central Park. Summer attracts nesting birds to the area and hummingbirds to the abundant jewelweed. It can be worth a trip to the outer reaches of Manhattan to bird this lovely and historic part of the city.

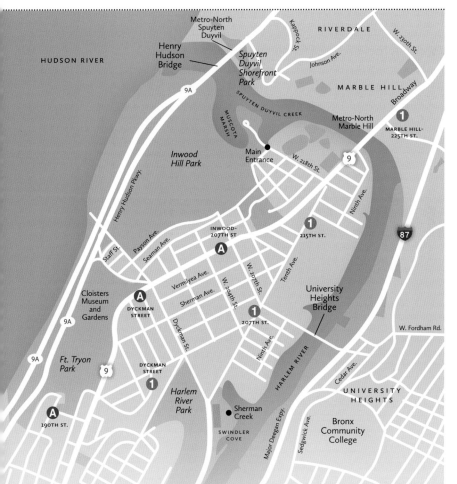

There are many ways to enter the park, but if you want to make the most of your time, focus on the northern part. Plan to visit the areas of Muscota Marsh, the paths along the eastern side of the ridge, and do a little exploring on your own, and you should leave this hot spot a satisfied customer.

Enter the park at the entrance at 218th Street and Indian Road, which will give you quicker access to most of the best birding areas. Continue along 218th Street into the park, and walk along the water. This is *Muscota Marsh*—the largest remaining natural saltwater marsh in Manhattan. In winter, you may find some ducks close in, and during migration, shorebirds may be poking around in the mud. Occasional rare species take advantage of this mucky area (Pectoral and Spotted Sandpipers, for example), but it's shorebirds like Least and Semipalmated Sandpipers you are most likely to see. To increase your chances, make sure you arrive at or near low tide, when there is plenty of exposed mud.

The nearby soccer fields may have interesting sparrows in fall. Make a left at the end of the soccer field and find Shorakkopoch Rock and the path leading from it heading north. Make a right on the path and shortly after check out the jewelweed for hummingbirds in summer.

Walk along this path as it gains elevation along the Spuyten Duyvil, and when you get to the base of the Henry Hudson Bridge make a left without passing underneath the bridge. These *wooded areas along the eastern ridge* in spring through fall harbor songbirds and resident breeding birds. Over two

Pectoral Sandpiper at Muscota Marsh

dozen species of warbler put down in Inwood Hill Park during migration, and breeding birds in summer include Eastern Towhee and Baltimore Oriole. But keep an eye out in winter and fall for Bald Eagles or migrating hawks in fall at this overlook. You'll be much closer to the birds at this elevation and may have some excellent views of birds you might only glimpse otherwise.

There is a charming old stone wall on this route, and great views and interesting trails to explore. If you have plenty of time, backtrack and pass beneath the bridge to take a walk around the peninsula, and scan out into Spuyten Duyvil and the Hudson River, where you will most likely see a lot of Ring-billed and Herring Gulls. This route takes you across and close to the highways that cut through the park and is less of a wilderness experience. A couple of paths that take you to different elevations reward you with amazing views of the river and the Henry Hudson Bridge—especially from the path closest to the water. In winter, this is the best place to look for Bald Eagles.

Our route is just a suggestion and covers some highlights. But a random wander through Inwood Park might offer interesting and serendipitous bird sightings. If you are open to a little exploration, don't pass up an attractive-looking pathway—some are not paved and less well traveled. These quieter routes are good for finding birds; however, beware of poison ivy encroaching upon the trail. Overall this is an attractive and productive place to bird.

Fort Tryon Park and the Cloisters Museum are just south of Inwood Hill. To reach these, exit on Payson Avenue and make a right. Enter Fort Tryon on Riverside Drive.

If you are combining this trip with *Swindler Cove* and *Sherman Creek*, you can walk from Inwood Hill to these parks via Dyckman Street, which is at the southern tip of the park. You will find fuller descriptions of these locations later in this chapter.

KEY SPECIES BY SEASON

Spring Great Crested Flycatcher; Blue-headed, Warbling, and Red-eyed Vireos; Northern Rough-winged, Tree, and Barn Swallows; Wood Thrush. Over two dozen species of warbler have been recorded here, including Blackpoll, Blackburnian, Wilson's, Black-throated Green, Nashville, and Blue-winged.

Summer Great Egret; Semipalmated and Least Sandpipers, and occasional Spotted Sandpiper; Pectoral Sandpiper have been seen; Chimney Swift, Eastern Screech Owls, Yellow-bellied Sapsucker, Eastern Wood-Pewee, Eastern Towhee, Wood Thrush, Brown Thrasher, Song Sparrow, Rose-breasted Grosbeak, Orchard and Baltimore Orioles, House Finch, American Goldfinch, Ruby-throated Hummingbird.

Fall Great Blue Heron and Black-crowned Night-Heron; Osprey, Sharp-shinned and Red-shouldered Hawks, Peregrine Falcon, Merlin, Bald Eagle; Spotted, Least, and Semipalmated Sandpipers; Laughing Gull; Ruby-throated Hummingbird, Belted Kingfisher, Ruby and Golden-crowned Kinglets, Brown Creeper, Hermit Thrush, Cedar Waxwings, Swamp and White-throated Sparrows.

Winter Waterfowl including Canvasback, Greater Scaup, Red-breasted Merganser, American Black Duck, Mallard; Bald Eagle; Red-bellied, Hairy, and Downy Woodpeckers; Red- and White-breasted Nuthatches, Black-capped Chickadee, Tufted Titmouse.

HABITAT

Woodlands, salt marsh, mudflats, edges.

BEST TIME TO GO

Year-round, but spring through fall is recommended.

HOW TO GET TO INWOOD HILL PARK

By Subway Ⓐ to 207th Street or ❶ to 215th Street.

By Car Take Broadway north to 218th Street, make a left, and the park is four blocks in. Street parking.

INFORMATION

Inwood Hill Nature Center, 218th Street and Indian Road

212-304-2365

http://www.nycgovparks.org

For information on the tides at Muscota Marsh, go to http://tides.mobile geographics.com and search Spuyten Duyvil for the Spuyten Duyvil Creek Entrance, Hudson River.

The wild areas are fairly wild, but this is a city park with many kinds of activities available, including picnicking, BBQ, dog runs, playgrounds, and athletic fields, which can attract thousands of visitors to the park. As you might expect, most human activity takes place around the playing fields.

Birding in this park will take you into some very remote and densely wooded areas, so we highly recommend going with a small group or, at the very least, one other person. It is not recommended that you explore here alone.

There are restrooms at the Urban Ecology Center on the Muscota Marsh peninsula, at the Isham Street entrance, and at the southern entrance at the Payson Playground.

No café, but there are a few food carts in the park.

Scope Yes, if you are going to Muscota Marsh to look for waterfowl or shorebirds. If you stick to the forest, binos should do the trick.

Photos Medium difficulty. Shorebirds can be close at Muscota Marsh, but when the surrounding vegetation is leafed out, it can be difficult to get a good shot. Likewise, the angle of light and time of tide don't always work together, so a little planning may be necessary.

Fort Tryon Park

Designed by Frederick Law Olmsted's son and donated to the City of New York by John D. Rockefeller Jr., Fort Tryon Park is an elegant woodland with beautiful stone bridges, lovely gardens, leafy pathways, and breathtaking views over the Hudson. Some of the property is contiguous with Inwood Hill Park, and similar woodland birds are found in both—there are just fewer of them in Fort Tryon. The Metropolitan Museum of Art's Cloisters Museum is at the north end, and if the birds are not cooperating, the art will more than satisfy.

This is a great choice if you are visiting the Cloisters and don't want to hike through Inwood. Lots of trees, planted gardens, and fine pathways offer respite for birds as well as humans. No scope is needed here unless you want to see the other side of the Hudson.

HOW TO GET TO FORT TRYON PARK

By Subway The Ⓐ stops at Dyckman Street, in the northeast corner of the park.

By Bus Take the M4 or M98 north to its last stop, which will be marked either 190th Street, Cloisters, or Margaret Corbin Circle. As you get on the bus, ask the driver if it goes to 190th Street, as not every one does. If it does, you'll be at the south entrance.

By Car There's no street address for the park itself, but you could use the address for the Cloisters Museum, 99 Margaret Corbin Circle, in your GPS. Drive up Fort Washington Avenue. At the traffic circle at 190th Street, it becomes Margaret Corbin Drive. Very limited parking is available along this drive near the café and by the Cloisters.

New York City Department of Parks and Recreation

212-639-9675

http://www.nycgovparks.org also provides a map

Fort Tryon Park Trust, 212-795-1388 x300, https://www.forttryonparktrust
.org

Information from the New York-New Jersey Trail Conference at
http://www.nynjtc.org

Fort Tryon has restrooms and a café.

OTHER THAN BIRDING

Trail hiking, gardens, playground, and of course the Cloisters. Rangers offer guided tours of the park (as well as occasional bird walks). Check the Fort Tryon Park and Fort Tryon Park Trust sites for more information.

Sherman Creek and Swindler Cove

Once a spot known for illegal dumping, these five acres on the Harlem River are being transformed. Now you can visit lovely gardens, a boathouse, and the centerpiece of the New York Restoration Project's work, Swindler Cove, whose native plantings and wetlands are attractive to a variety of birds. Enjoy the views out onto the Harlem River, the migrating fall shorebirds, and the common waterfowl that seek refuge here in the winter. The birding here is good but not great, although this is one of the only places in Manhattan where peeps are regularly seen. These parks are intimate and have a variety of diminutive habitats. They are not as well-birded as the nearby and much larger Inwood. This can be a pleasant outing with non-birding friends, who will enjoy the gardens while you look for songbirds and scan the shoreline for waders. Best results will come during low tide, but for more serious shorebirding you should check out Muscota Marsh in Inwood.

VIEWING SPOTS

Enter *Sherman Creek* from Tenth Avenue and take the path toward the water. The gardens are very nice, but the more birdy spots may be toward the water, where it's a little wilder. Check for migrating shorebirds by the river's edge, and on the Harlem River for waterfowl in winter.

The path soon takes you to *Swindler Cove*. This is a delightful park with a variety of interesting areas and tiny habitats that might be a little more productive than Sherman Creek. Its freshwater garden may harbor some wrens, and it has nesting gourds by the restrooms and a nice wooded area. Stroll around here during migration and look for Yellow Warbler, Common Yellowthroat, and American Redstarts. Song Sparrows, Red-winged Blackbirds, and backyard birds are found throughout the park, as are woodpeckers closer to Dyckman Street. If you didn't get to see enough birds and you want to make Inwood Park your next stop, you can get there by walking west on Dyckman, crossing Broadway, and making another right on Payson into the park.

If you want to combine good birding with a world-class medieval art experience, the Metropolitan Museum of Art's Cloisters Museum is on your left in Fort Tryon Park as you cross Broadway, and makes for a nice secondary stop when visiting Inwood Hill Park.

KEY SPECIES BY SEASON

Spring Yellow Warbler, Blackpoll, American Redstart, Common Yellowthroat; Red-tailed Hawk, Black-crowned Night-Heron.

Summer Nesting birds include Eastern Kingbird, Warbling Vireo, Barn Swallow, Song Sparrow, Red-winged Blackbird, Baltimore Oriole, American Goldfinch, House Finch.

Fall Songbirds; shorebirds such as Killdeer, Spotted, Solitary, Semipalmated, and Least Sandpipers, Greater Yellowlegs; woodpeckers.

Winter Canada Geese and Mallards in great numbers, American Black Duck, occasionally a Wood Duck, Gadwall, or Ruddy Duck.

HABITAT

Shoreline, edge, gardens, forest, wetlands.

BEST TIME TO GO

Year-round, but late spring through fall has the most variety of birds.

HOW TO GET TO SHERMAN CREEK

The entrance to Sherman Creek is at Tenth Avenue between Academy Street and the Harlem River. The ❶ stops at Dyckman, the closest subway station. From there it's just a few minutes' walk straight on Tenth Avenue to the entrance.

Sherman Creek
https://www.nyrp.org
http://www.nycgovparks.org

Swindler Cove
https://www.nyrp.org
No café, but restrooms are available in Swindler Cove.

OTHER THAN BIRDING

There is an established community garden; a kayak launch is in the works.
Scope Not necessary.
Photos Easy.

Randall's Island

If you want a change of scene and don't want to travel too far, explore this island in the middle of the East River. Randall's is a remarkable area with recreational fields and open spaces, beautiful shorelines, great views, and well-maintained natural areas.

Once separate, Randall's and Wards Islands are now known simply as Randall's Island. They have been part of New York City's history since 1637, when they were purchased from Native Americans by the Dutch. Little Hell Channel, which separated the two, was filled in during the twentieth century. Over time, the island has undergone an enormous transformation from an urban out-island to a public park with plenty of space for recreational uses. Thankfully, some of it has been set aside for wildlife. Birds and other animals find refuge in nine acres of carefully managed fresh- and saltwater marshes and native plant areas. All of this takes place beneath a major highway and elevated train bridge.

There is a lot of ground to cover on this 432-acre island, but it can be easily traveled by bike, by car, or with several hours of walking. A shortened trip, indicated on the map, will take you to the most productive areas.

VIEWING SPOTS

How you approach and arrive at Randall's Island will determine where you start. If you're walking from Manhattan, the footbridge offers the most direct route. From the bridge, stop on the way over and look to the right. You will see a small island in the middle of the harbor called *Mill Rock*. This island, closed

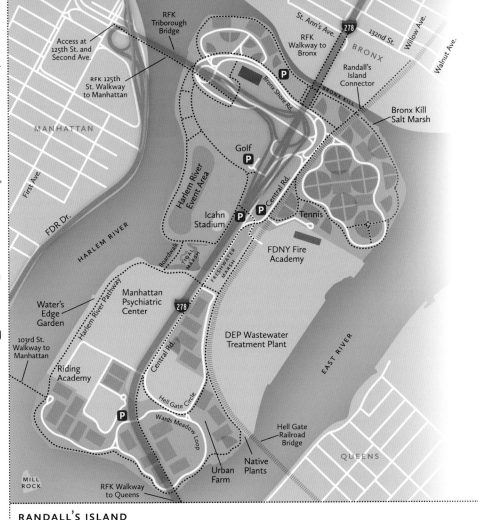

RANDALL'S ISLAND

to the public, has been permitted to return to its natural state. On it you may find nesting Great Egrets, Snowy Egrets, Black-crowned Night-Herons, gulls, and cormorants.

Once across the bridge, if you want to take a walk with stunning views of Manhattan, turn right. If you want to start birding immediately, turn left and follow the Harlem River Pathway along the water and look in the river for cormorants and, in late fall and winter, waterfowl. Check out the beautifully maintained *Water's Edge Gardens* for sparrows, warblers, vireos, and finches during migration; in winter you may find Red-bellied and Downy Woodpeckers, along with Tufted Titmice, Black-capped Chickadees, and Yellow-bellied Sapsuckers.

MANHATTAN

Kestrels and Peregrines are seen on or near the Manhattan Psychiatric Center year-round.

Continue on this path and check out the trees at the north end, just past the bridge, then walk back and turn onto the bridge leading into the wetlands. This tidal marsh can have a variety of sparrows, including Swamp Sparrows, as well as American Black Ducks and, in winter, Gadwalls. Also look for herons in the marsh and cormorants in the river. Once you have walked over the boardwalk, make a right to walk under the highway and cross Central Road.

Find the entrance to the *natural area* near the driveway to the wastewater treatment plant at Hells Gate Circle. It is clearly marked and can be a highlight of your trip to Randall's Island, as this wetland is home to a variety of birds, including nesting Yellow Warblers, Eastern Kingbirds, and vireos flitting through the tiny wilderness. There is also a fearless muskrat. The paths are lined with native plants designed to attract birds and butterflies. Walk back out to Central Road and bird the trees around the wetlands from the road.

For those with plenty of time and energy, make a right toward the New York Fire Department Training Division (in fact, you may see new recruits training in the park), and heading north, scan the *ball fields* for Barn and Rough-winged Swallows. This can also be a good place to see a variety of sparrows, including White-throated, Chipping, Vesper, Song, and even White-crowned.

Follow Central Road past *Icahn Stadium* and check to see if the Red-tailed Hawk nest is still active on the stadium light post behind the fields. At the footbridge to the Bronx, make a right and walk the coast along the *Bronx Kill*. Gulls and herons may be found here, as well as Gadwall, American Black Duck, Common Loon, Red-breasted Merganser, and Killdeer. Swallows can also been seen working the fields. Nelson's Sparrow have been reported.

Now make your way west to visit the *northwest corner* of the park. A long-term construction project has been going on here, so not all of it may be accessible. Herons and egrets can be found off the Harlem River Parkway toward the north of the island.

A trip to the southern portion of the island, formerly Wards Island, can be productive, although the middle and north, described above, have the best natural areas and most potential. An urban farm to the east, south of the wastewater treatment plant, can be reached by following Hell Gate Circle and making a right onto Wards Meadow Loop. Follow this road to the water and you'll find yourself between the Hell Gate Railroad Bridge and the RFK Triborough Bridge in a surprisingly tranquil *native planting area* with benches to enjoy the view of Astoria Park across the river. This area can be birdy and offers a welcome breezy respite in summer.

Strolling by the water may be more about stunning views of Manhattan and Mill Rock than of birds, but what spectacular views they are! Look for cormorants and the occasional waterfowl in the river. If you have time, it's worth the effort, and leads you back to the 103rd Street footbridge.

KEY SPECIES BY SEASON

Spring Some of the interesting waterfowl persist through early spring. Great and Snowy Egrets; shorebirds including Killdeer, Spotted Sandpiper; some songbirds, especially in the wetlands.

Summer Yellow Warbler; Song and Savannah Sparrows; Red-winged Blackbird; swifts and swallows, including Chimney Swifts and Barn and Tree Swallows; Baltimore Oriole.

Fall Some migrating raptors; returning songbirds, including White-crowned, Swamp, Nelson's, and Saltmarsh Sparrows; Red-bellied and Downy Woodpeckers, plus Northern Flicker and Yellow-bellied Sapsuckers through spring.

Winter The usual overwintering songbirds and waterfowl, plus a chance to see Red-throated Loon, Common Loon, Red-necked Grebe, Great Cormorant, Common Goldeneye, Gadwall, Iceland Gull. You could also find hawks, American Kestrel, and Peregrine Falcon; Song, Swamp, and White-throated Sparrows. Horned Lark, Snow Bunting, and American Pipit are possible. American Black Duck can be found year-round.

HABITAT

Freshwater and tidal wetlands, a small urban forest, and four and half miles of coastal upland habitat.

BEST TIME TO GO

Year-round.

HOW TO GET TO RANDALL'S ISLAND

By Subway and Bus The M35 bus runs to the island from the northwest corner of 125th Street and Lexington Avenue. Transfer is available from the Lexington Avenue ❹ ❺ or ❻ at 125th Street.

By Foot or Bicycle Randall's Island Park offers approximately eight miles of pedestrian and bicycle pathways, accessible from points in Manhattan, the

Bronx, and Queens. There is no bicycle rental on the island itself. The 103rd Street Footbridge in Manhattan is open to pedestrians and cyclists year-round and is accessible at the East River Esplanade at 103rd Street and FDR Drive. Take the ⑥ to 103rd Street or the M15 bus to either 100th or 102nd Streets. Walk east along 102nd Street to FDR Drive. Then walk one block north directly onto the crossover leading to the footbridge. Alternatively, take the M106 across town to FDR Drive and walk three blocks south. On Randall's Island, the footbridge is accessible at the southwest corner of the Island.

Pedestrian walkways on all three spans of the RFK Triborough Bridge connect the park to Manhattan, the Bronx, and Queens. The better route from the Bronx, originating at 132nd Street between Willow and Walnut, that takes you to Randall's underneath the train bridge to Central Road, is the Randall's Island Connector, which is at grade—no schlepping up and down the Triborough's Staircases.

By Car All automobiles enter the island via the RFK Triborough Bridge and will have to pay a bridge toll. If navigating with GPS, use "20 Randall's Island Park, New York, NY 10035" as your destination.

INFORMATION

http://randallsisland.org
http://www.nycgovparks.org

There are restrooms throughout the island. Many are not accessible all the time, but look for the orange pods that are scattered throughout. These restrooms are always open and are clean.

Food is hit or miss here as well. The official map shows several places to get food, but they are not always open. It might be best to bring your own lunch if you plan to stay much of the day. One tip to score lunch: if there is construction going on, look nearby for a food truck, where you should find a serviceable, affordable meal and friendly service. If Icahn Stadium is open for a track meet, you might be able to go inside and buy a drink and snack at the concession upstairs.

The Randall's Island Park Alliance keeps the natural areas in a state attractive to birds and wildlife as well as people.

OTHER THAN BIRDING

A huge range of sports facilities for track, mini-golf, tennis, baseball, soccer, and cricket are available here. Gardeners will be entranced by the plantings in the natural areas and along the Harlem River. Concerts take place here in the warm months. A yearly event in May, the Frieze Art Fair, brings thousands of people to the island at the peak of migration, so check the events calendar before you go.

Scope You might yearn for it to take in views of the birds on Mill Rock, but if you're walking, consider how far you really want to carry it.

Photos Easy.

Governors Island

Take a step back into New York City's history with a short ferry ride to Governors Island. Thanks to its location in the middle of New York Harbor, for two hundred years it was an active military base. Now decommissioned and open to the public, it has approximately two miles of shoreline, 172 acres of open and wooded areas, and retains some of the buildings from its army and coast guard days. It also offers a staggering view of the Statue of Liberty. The ambitious plan to convert the island to a tourist destination is still under way as of this writing, and that includes adding more fields and planted areas, which should increase its attractiveness to birds. Currently, the island is open only in the summer, and it will not be considered a prime spot for birding until visitors are allowed during spring and fall. It does, however, host a summer breeding tern colony (which alone may be worth the visit) and has enough other stuff going on for an enjoyable outing with non-birders.

VIEWING SPOTS

Head to *Yankee Pier* for the Common Tern colony. June and July is the best time to see the chicks. On weekends this pier is always open. During the week there is limited or no access, but the terns and cormorants on the pier may be seen from the entrance to the Brooklyn ferry or Himmel Road, which runs along the water by the pier. Scan the rocks for shorebirds, and the rest of the area for gulls and the occasional pelagic. Ship traffic throws a lot of wake on the rocks, so sightings may not be dependable.

Common birds in *the park* include a variety of woodpeckers, like Northern Flickers, and the usual backyard birds. The *new garden area* behind the food courts and playing fields may produce Song Sparrows and a variety of swallows. Red-tailed Hawks and American Kestrels are seen. Yellow-crowned Night-Herons sometimes nest here.

A walk to the *east side of the playing fields* will reveal the best views of the Statue of Liberty. En route you may see swallows working the fields, gulls, and some shorebirds overhead.

When the island opens every year in May, you can expect migrating songbirds like Northern Parula, Common Yellowthroat, American Redstart; Yellow, Yellow-rumped, Blackpoll, Black-throated Blue, and other warbler species blast through. In summer, mostly seen are the typical backyard birds of southeastern New York. Common Terns nest here, as described above. Also, you may see cormorants; Red-eyed Vireo; Chimney Swifts; Northern Rough-winged, Barn, and Tree Swallows; and Chipping and Song Sparrows. The park is closed in winter, but private boats circling the island will get looks at winter waterfowl such as Red-Breasted Mergansers, Buffleheads, Brant, Greater and Lesser Scaup, and Gadwall.

The park is easily walkable, but renting bikes or bringing your own on the ferry is a great way to get around. Governors Island is really a compromise for birders who have non-birders with them. Nevertheless, this is a fun outing and a great place to walk, bike, and bird. Taking photos is easy, but leave the scope at home.

HOW TO GET TO GOVERNORS ISLAND

The island is accessible only by ferry, from either Manhattan or Brooklyn. The Manhattan ferry leaves from the Battery Maritime Building at 10 South Street on the corner of South and Whitehall Streets in lower Manhattan—next door to the Staten Island ferry terminal. The Brooklyn ferry leaves from Brooklyn Bridge Park's East River Pier 6 at the end of Atlantic Avenue.

Commercial ferry service is provided by the East River Ferry from points in Manhattan, Greenpoint, Williamsburg, Fulton Ferry, and Wall Street; www.eastriverferry.com.

INFORMATION

http://www.nps.gov/gois. Bird sightings from prior years can also be
    found at this website.
http://www.govisland.com

Construction of the new park area is expected to continue for several years. As a result, the number and species of birds in the interior of the island may change. The bird census, which was done for several years before the con-

Common Tern

struction commenced, might shed some light on which birds can be expected at this location once work is complete. However, a total transformation of the southern portion of the island is in the works. The introduction of gardens and fields will hopefully make Governors Island a more attractive location for a wider variety of birds.

Restrooms are located throughout the island. You'll also find benches, playgrounds, and plenty of places to buy food, including the food court, where you can sit outside and watch swallows sail over the flower beds. Facilities are closed on Mondays and Tuesdays.

OTHER THAN BIRDING

If you are traveling with children or non-birders, Governors has lots of activities to keep them happy. The National Park Service offers several programs dealing with the military history of the island, such as tours of Castle Williams, the best-preserved circular fortification in the United States. There are ongoing art programs for kids and a variety of cultural and artistic programs and events throughout the season.

Hudson River Greenway Biking

Grab your bike and binos for a birding tour of Manhattan's West Side. It's not necessarily a spin through the best hot spots, but it's a great ride, and the

Hudson River often delivers some interesting waterfowl. During migration, you can investigate some of the parks as you pass by. As you travel along, you will likely see raptors and, in the northern section, resident Monk Parakeets. This fun outing takes you all the way from Fort Tryon Park in the north to the Battery at the southern tip. Along the way you pass through Fort Washington Park, Hudson River Park, and Riverside Park.

Non-birders in your group will find interesting sights along the way: Grant's Tomb, the historic Little Red Lighthouse under the George Washington Bridge, the Cloisters Museum in Fort Tryon Park, the USS *Intrepid*, and the Battery and its neighborhood museums.

VIEWING SPOTS

The Hudson Park Greenway starts in Fort Tryon Park, just below Inwood Hill Park, and runs south to the Battery, from where it then goes up the East Side, connecting with the East River Greenway (see our descriptions of Inwood Hill Park and the Battery in this chapter). If you take this route you will pedal nearly twelve miles one way over mostly level terrain.

Our tour takes you north to south, but you can do it either way. Since you're likely to see more birds in the northern section, we start there. Moving south from *Fort Tryon Park*, look for waterfowl in the Hudson late fall through early spring until you reach the water treatment plant on 145th Street. It's not a certainty, but along this stretch there can be a variety of birds, including mergansers, Peregrine Falcons, Monk Parakeets, and the occasional Bald Eagle.

Continue toward the Fairway grocery store at 130th Street and, in winter, check before you get to the store for Canvasbacks and Bufflehead just north of it. Keep an eye out year-round for raptors overhead.

Head south through *Riverside Park* and take time to leave the shore of the Hudson and pedal around inside the park (see our description of it in Other Places to Find Birds in Manhattan). After you've made it through Riverside, you will have seen most of the birds you will likely encounter for a while. South of 59th Street, things get more commercial, with cruise ship piers and the battleship *Intrepid* museum—all making for a less-than-birdy experience—although it's a good urban bike ride with a number of tourist attractions along the way.

Things start to pick up again from a birding perspective as you approach the Meat Packing District around 14th Street. Below 14th, start looking in the water when you reach Christopher Street. Brant love this area, although they roam around and may be nearer Chelsea or farther south. As you reach Hudson River Park Pier 40, you may see some Mallards and other ducks on the north side, but the south side is more protected and affords a better chance for waterfowl.

The vent to the Holland Tunnel's entrance is *just south of Pier 40*, where Spring Street and Canal Street converge. Look for the little secluded area of pilings that provide a temporary home to migrating Bufflehead, Red-throated Loons, and other waterfowl that dive among and nibble on the pilings. You know you've arrived at your destination when you see two piers side by side . . . and one fully locked down. Just south of the locked gate between the two piers is this little spot, and it's worth swinging by if you are in the area. By going out on the pier that is open to the public, you'll get great views of the Statue of Liberty and passing ocean liners. In late summer, Common Terns often nest just below the railings on the south side. You can see the nests and the birds hanging out. Also look at what remains of a jetty just south of the pier, as there are sometimes cormorants and interesting waterfowl and gulls there.

Continue on to *Battery Park City*, which is just below Harrison Street. On your way, to the east of the bike path, there is a planted area that may have some activity during migration. At Harrison Street, make a right to stay alongside the water as you cruise through Battery Park City and past the water taxi pier. The Nelson A. Rockefeller Park in front of the apartment buildings may also have some interesting sparrows and warblers during migration.

Keep going south and you will eventually arrive at *the Battery*. See our description in this chapter. This is a nice outing with friends or to see the city by bike and do some casual birding, but you can do better if you just want to see birds in Manhattan.

INFORMATION

http://www.hudsonriverpark.org

If you need to rent a bike, there are a number of rental locations along the route:

CitiBike is open year-round. This bike-sharing operation has the most locations near the Hudson River Greenway. http://www.citibikenyc.com.

Bike and Roll bike rental locations may close in winter. Please check before heading out to be sure they are open. http://bikenewyorkcity.com.

Restroom Locations (some are closed in winter):
  In Fort Tryon and Riverside Parks
  Clinton Cove at West 55th Street (portable toilet)
  Pier 84 at West 44th Street
  Chelsea Waterside at West 24th Street (portable toilet)
  Pier 51 at West 12th Street
  Pier 45 at Christopher Street
  Pier 40 at West Houston Street
  Pier 25 at North Moore Street

The city has done a remarkable job making the West Side appealing to bikers and has added many new bike paths, connecting Inwood Hill Park in the north to the Battery in the south, with a nearly continuous trail up the East Side, so that you can walk or bike the entire city. However, please note that some areas—especially in the 40s—require navigating city traffic. While traffic lights make it more manageable, if you are not accustomed to urban congestion, you may not feel comfortable. If that is the case, biking the northern area above 59th Street, or the area below 34th Street, will be a safer and more rewarding ride.

The Battery

This beautifully planted and maintained European-style park with a striking promenade sits at the southernmost tip of Manhattan facing New York Harbor at the confluence of the Hudson and East Rivers. The area was settled by the Dutch in 1623 because of its strategic location, and the first "battery" of cannons was set up here to defend the new settlement. Beginning in 1855, this was the first stop for immigrants entering the United States. Now it's a popular tourist attraction, with a number of historical sites, fabulous vistas on the water, and a lot of things to do.

The spots to check out are the Battery Woodlands, the urban garden and Forest Farm; across from Castle Clinton there are some fruit trees, which explode with flowers in spring and are attractive to songbirds. Be sure to check out the harbor before walking east toward the apiaries. Despite heavy ship traffic, a rare bird may be bobbing around.

From here in summertime you can jump on a ferry to *Governors Island* and find the nesting Common Terns—see our description of Governors Island in this chapter. You can also combine this with a visit to Wall Street, the National Museum of the American Indian, and other popular tourist spots. If you are riding a bike, you can also now continue around the southern tip of Manhattan through the Battery and go up the East Side, although this side of Manhattan is much less birdy.

Even though this is not a prime hot spot, it's a good place for both birders and non-birders—especially in spring when songbirds feed among the flowers.

HOW TO GET TO THE BATTERY

The Battery is easily accessed by public transport, as well as bike via the Hudson River Park Greenway (see our description earlier in the chapter).

By Subway The closest stops for the southernmost location, which also offers access to the Staten Island Ferry, are South Ferry on the ❶; Bowling Green on the ❹ or ❺; Whitehall Street on the Ⓡ.

By Bus The M9, M10, and M22 buses all stop in Battery Park City.

By Ferry The Trans-Hudson Ferry runs directly from Hoboken and Jersey City to the Nelson A. Rockefeller Park in Battery Park City. And, of course, there's the Staten Island Ferry.

INFORMATION

http://www.nycgovparks.org
http://www.thebattery.org
http://www.nyharborparks.org

There are plenty of restaurants inside as well as just outside the park. Restrooms are available.

The Battery has been undergoing a major transformation for many years and as of this writing is nearing completion. Attention has been paid to the use of native plants, urban farming, and sustainability, and as a result this busy park should become even more attractive to migrant birds.

OTHER PLACES TO FIND BIRDS IN MANHATTAN

Having birded Central Park, you'll realize that these other places are really just smaller, less-productive versions. Not destinations per se, these are more a way to get a quick migration fix when on a work break, out with friends, or when you're in the area and have some time to kill. As a bonus, many are in interesting or historic neighborhoods, and before you know it, you may be packing binos with you wherever you go . . . just in case. The most productive one, Bryant Park, is marked with an asterisk.

Bryant Park*

It's probably the least-likely looking hot spot in New York, but it does have surprisingly good birds during migration. Of all the parks in this section, and despite its constant stream of visitors and winter fairs, it is by far the most highly recommended. Located between Fifth and Sixth Avenues and between 40th and 42nd Streets, it's a busy park directly behind the New York Public Library and easily accessed by several subway and bus lines. Thanks to sharp-eyed frequenters to this park, unusual birds are sometimes reported. American Woodcocks, for example, often are found in the planted areas behind the library during spring migration. Otherwise, count on a subset of the birds you'd find on the same day in Central Park. The (B) (D) (F) (M) (7) subways are closest, but it's also near (N) (Q) (R) (1) (2) (3) (4) (5) (6) (S). It is also reachable by the M5, M7, and M42 buses and is near Grand Central Terminal and the Port Authority Bus Terminal. If the birds aren't cooperating, there are great restaurants in and around the park and, of course, the splendid New York Public Library Building itself. Bryant Park has public restrooms open year-round.

Madison Square Park

This carefully maintained park located in the Flatiron District and bordered by Madison Avenue and Fifth Avenue and 23rd and 26th Streets doesn't really offer much to attract birds. However, during spring and fall, warblers and sparrows sometimes find their way here, so if you are nearby you might want to take a look. This is another park that is surrounded by enough dining and shopping to compensate a bit if the birds are sparse. Easily reached by the (N) (R) (6) and PATH train, or by the M1, M2, M3, M5, or M23 buses.

Union Square Park

Union Square Park is also easy to get to by public transportation and is in a location convenient to lower Manhattan. On certain days of the week

American Woodcock in Bryant Park

there's a world-class green market, so it's a nice chance to combine birding and sustainable grocery shopping. You will be most likely to see American Kestrels and Red-tailed Hawks, although warblers and sparrows migrate through in spring and fall. In winter, you'll see Yellow-bellied Sapsuckers and White-throated Sparrows. The park is between 14th and 17th Streets, and between Park Avenue (where it merges with Broadway) and Union Square West, before it becomes University Place. The 4 5 6 L N Q R subways are nearest, but the F M 1 2 3 work as well. By bus, take the M1, M2, M3, M14a, M14d. PATH trains are close, too.

Washington Square Park

It's convenient to public transport and the New York University campus and the West Village. You'll have a good chance of seeing Red-tailed Hawks year-round, and maybe some warblers and other songbirds during migration. Bounded by University Place, Waverly Place, MacDougal Street, and East 4th Street. By subway A C E B D F M N R 1 6; by bus M1, M2, M3, M5, M8.

Mourning Dove

Morningside Park

This long slice of a park is close by Columbia University. It occasionally traps a migrant songbird or two and in winter has resident nuthatches, chickadees, and some sparrows. The pair of Red-tails nesting on St. John the Divine make spot appearances. Just east of Amsterdam Avenue, the park is nestled under a cliff and bounded by West 123rd Street, Morningside Avenue, West 110th Street, and Morningside Drive. The ❶ Ⓑ Ⓒ subways are closest, but you could easily get there on the Ⓐ Ⓓ ❷ ❸ or by the M3 or M4 buses.

Riverside Park and "the Drip"

This scenic four-mile stretch of beautifully designed park by Fredrick Law Olmsted is a joy to bikers and walkers and has a spectacular waterfront on the Hudson River. While the park is officially designated a scenic landmark in the City of New York, most of it lacks appeal to birds, except for the well-known water source called the Drip. This is a sit spot, although you'll have to bring your own chair. During migration, a spigot is turned on, and birds come down to drink and bathe after foraging in the surrounding trees. It's difficult to find your first time, but if you enter at 116th Street and Riverside Drive and then head north toward the tennis courts, you'll see a small fenced-in area and perhaps a birder or two patiently waiting for that special warbler. Take the ❶ or the M5.

Carl Schurz Park and the East River

Carl Schurz is a friendly neighborhood park next to the FDR Drive and East River that was designed by Frederick Law Olmsted. It can get an occasional migrant and offers a convenient walkway out onto the river, where, in winter, waterfowl may be hanging around. If you are visiting Gracie Mansion, take a stroll along the water and check out the trees for avian activity. Enter the park at East 87th Street and East End Avenue. Once the Second Avenue subway is running, it will provide the closest stop. Until then, you'll have to walk over from the ❹❺❻ at 86th and Lexington Avenue. The closest buses are the M31 and M86.

Peter Detmold Park

This secluded park on the East River not far from the United Nations is reached by a rather forbidding staircase from either 49th Street or 51st Street and FDR Drive. The park tumbles down the side of a cliff overlooking the East River and has a popular dog run. Detmold features trees and a walkway to an overlook on the East River from 51st Street where Red-breasted Mergansers and Brant may be around in early spring. Nearby Peregrines and Red-tailed Hawks sometimes are seen over the East River. The Second Avenue Subway, when completed, will make it an easier place to visit. Until then, the closest subway stops are the 51st Street stop for the ❹ and ❻ and the Lexington Avenue / 53rd Street stop for the ❻ and Ⓜ. By bus, the M15 and M50 are closest.

UNIQUELY MANHATTAN BIRDING

New York offers some interesting birding that is available only in Manhattan. Try one of these to add some variety to your outings:

Take the Water Taxi to Swinburne and Hoffman Islands

These two man-made islands in New York Harbor were nineteenth-century quarantine stations for immigrants suspected of having contagious diseases. Now they are off-limits to people and are reserved for the benefit of wildlife—hosting Harbor Seals in winter and providing safe nesting for herons and other waterbirds in summer. Special tours operated by New York City Audubon or New York City Parks in summer and winter allow you to hop on the bright yellow water taxis and swing by for close-up views. In the summer, you can also take a worthwhile trip to *North and South Brother Islands*, the only way to see the birds nesting there.

Take a Bike Ride on the Hudson River Park Greenway

See our description earlier in this chapter.

Take a Tour of Peregrine Sites

There are quite a few of them, but these are generally the most reliable:

Brown Thrasher

55 Water Street—on the fourteenth floor. You can also see the nest from Brooklyn Bridge Park, but it's very far away.

Most bridges entering Manhattan have nesting Peregrines.

MetLife Building—200 Park Avenue at 45th Street—next to Grand Central. Stand on Park Avenue south of the building and point your binos at the MetLife "M."

Walk the High Line

This won't be the birdiest stroll you'll take in New York City, but it will be one of the more interesting ones. The conversion of a raised railroad line into an aerial greenway is another one of the West Side's success stories. A beautifully designed and maintained 1.4-mile-long path from Gansevoort Street (at Washington Street) north to 34th Street (west of Tenth Avenue) has a variety of views of the Hudson. While native plants are used, the overall idea is decorative, and not a lot of birds find them attractive. As a birding destination it's not ideal, but many people find themselves here having a wonderful time with friends. If you happen to be packing binos, you may catch a glimpse of something interesting. There are many events and tours here, including art installations and weekly stargazing. Check their website, http://www.thehighline.org/, for what's happening.

Look for Bald Eagles in Winter

Those iconic symbols of freedom and patriotism were nearly decimated, along with other birds of prey, in the 1960s because of DDT. Happily, they have made a roaring comeback and are beginning to appear around Manhattan. We've been seeing them for years at Inwood Hill Park, where a Bald Eagle reintroduction project was attempted, and they can be found from time to time soaring over New York Harbor. The latter sightings may have been a precursor to the first nesting pair in 2015. The exact location is kept a secret for their own safety, but they have set up housekeeping on an uninhabited island in the waters around Staten Island.

And if you can't leave home or the office, there is still hope:

Gadwall

Nest cams! Yes, there are nest cams for many of the nesting raptors around Manhattan. Here are a few that are current as of this writing:

Peregrine Falcons
At 55 Water Street: http://www.55water.com
At Tappan Zee Bridge: http://www.newnybridgegallery.com/falcon-cam.php

Red-tailed Hawks
At NYU Bobst Library: http://original.livestream.com/nyu_hawkcam
At New York Times: http://original.livestream.com/nytnestcam

2 Brooklyn

Named Breukelen by the early Dutch settlers, this western edge of Long Island is the most populated of the five boroughs and an amalgam of diverse birding hot spots rivaling anything found in the City of New York. The birding is loosely divided into parks and coastal areas, and the best of Brooklyn includes *Prospect Park*, a large urban Olmsted-and-Vaux-designed park that has been designated an Important Bird Area; the lovely *Brooklyn Botanic Garden*; historic *Green-Wood Cemetery*; and an old military airfield. The coastline, framed by New York Harbor and the Hudson River, is part of the Gateway National Recreation Area and offers urban coastal birding at its best.

Of the parks, Prospect is a premier destination, rivaling Central Park. In winter, *Brooklyn's coast* really shines as a birding destination for the volume and variety of waterfowl that can be seen in open water from the piers and shoreline. Straddling those two worlds is *Floyd Bennett Field*, a derelict airfield with 140 acres of precious grassland habitat and some forest, surrounded by the waters of Jamaica Bay.

KEY SITES

Prospect Park

After designing Central Park, Frederick Law Olmsted and Calvert Vaux spent thirty years creating this lovely 585-acre park nestled in the heart of Brooklyn. Less manicured than Central Park, it provides a more natural setting for viewing its 270 recorded species in man-made marshes, rolling meadows, and dense woodlands. This is Brooklyn's migration hot spot, although year-round there are plenty of birds to keep you busy and give you a reason to visit. The park is so attractive to birds and provides such critical habitat it has been designated an Important Bird Area.

NEW
JERSEY
JERSEY
CITY
MANHATTAN
HUDSON RIVER
EAST RIVER
GREENPOINT
495
Long Island Expy.
Woodhaven Blvd.

★ KEY SITES

478
Brooklyn
Bridge
278
BROOKLYN
Forest
Park

Ellis
Island

Statue
of
Liberty
Governors
Island
BROOKLYN
HEIGHTS
Brooklyn
Bridge
Park
Atlantic Ave.

BROOKLYN
QUEEN

RED
HOOK
PARK
SLOPE
Eastern Pkwy.
Linden Blvd.
Belt Pkwy.

UPPER NEW
YORK BAY
Bush Terminal
Piers Park
27
★ Prospect
Park
★ Brooklyn
Botanic
Garden
27

Brooklyn Army
Terminal Pier 4
Owls
Head
American
Veterans
Memorial
Pier
278
★ Green-Wood
Cemetery
FRESH
CREEK
BETTS
CREEK
HENDRIX
CREEK
Spring
Creek Pa

CANARSIE

Narrows
BAY
RIDGE
Flatbush Ave.
Ocean Pkwy.
Canarsie
Park
Canarsie
Pier

STATEN
ISLAND
GRAVESEND
278
JAMAICA
BAY

FT. WADSWORTH
Verrazano
Narrows
Bridge
Middle Parking Lot
● Dyker Beach
● South Parking Lot
Kings Hwy.
Marine
Park
GERRITSEN
CREEK
Ruffle
Bar

★ Floyd
Bennett
Field

Belt Parkway East

LOWER NEW
YORK BAY
Coney
Island
Creek
Park
SEAGATE
Calvert
Vaux Park
CONEY ISLAND
Plumb
Beach
★ Dead
Horse
Point
Jacob Riis
Park

Coney
Island
Pier
● Fort
Tilton

Breezy
Point

BROOKLYN

VIEWING SPOTS

Consult the map for the suggested route that takes you through the most worthwhile areas. Enter at Grand Army Plaza and follow the path through the wood and edge habitat known as the *Rose Garden* and the *Vale of Cashmere*, where there are lots of shrubs that can also be productive. The open area at *Nellie's Lawn* is often interesting for sparrows. *Long Meadow* is also birdy— check the trees along the edge, as well as the open space. During migration, if you visit early, you may see hawks as well as interesting grassland and ground-feeding birds in this area before the runners and dog walkers come out.

Ideal warbler habitat is found in *Midwood*. Spend time wandering around here during migration. At the north end of Midwood, *the Pools* are good places to search for waders. From here continue walking west through *Quaker Hill*

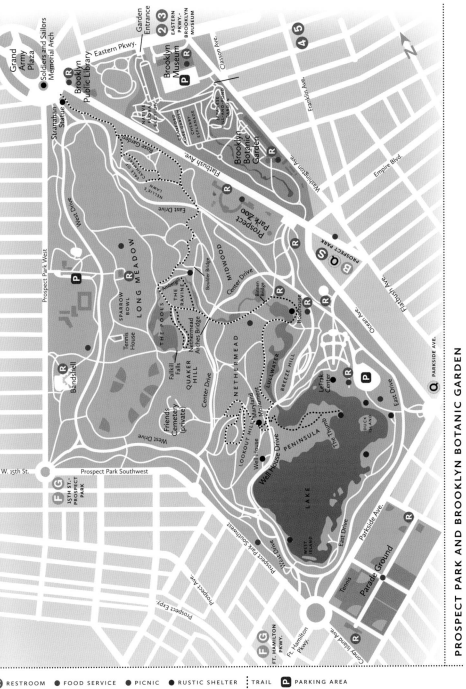

PROSPECT PARK AND BROOKLYN BOTANIC GARDEN

RESTROOM ● **FOOD SERVICE** ● **PICNIC** ● **RUSTIC SHELTER** ┊ **TRAIL** 🅿 **PARKING AREA**

and around the perimeter of the cemetery, which is warbler territory during migration. Another choice would be to take the route through *the Ravine* and *Nethermeade*, heading toward the Audubon Center in the handsome nineteenth-century *Boat House*.

From here, make your way through *Lullwater* and turn left onto *the Peninsula*, which is worth a thorough search. At the end of the Peninsula, check out the islands in *Prospect Lake* for nesting and roosting waterfowl and waders. From here walk up *Lookout Hill*, the highest point in the park. It has a reputation for warbler fallouts during migration and can be particularly productive.

KEY SPECIES BY SEASON

Spring Over three dozen species of warblers, including Blue-winged, Black-and-white, Bay-breasted, Black-throated Green; Eastern Wood Peewee, Eastern Phoebe, the occasional Acadian or Great Crested Flycatcher passing through, Brown Creeper. Rose-breasted Grosbeak.

Summer A nice diversity of nesting birds for an urban park. Less common species include White-eyed Vireo, Indigo Bunting, and Orchard Oriole. Green Heron, Wood Duck, Warbling Vireo, Wood Thrush, Brown Thrasher, Yellow Warbler, Eastern Kingbird, and Baltimore Oriole. Carolina and House Wrens also nest here.

Fall Warblers pass back through, plus a greater number and variety of sparrows, including Savannah, Fox, Swamp, White-crowned; Sharp-shinned and Cooper's Hawk, Merlin and the occasional Bald Eagle; Spotted Sandpiper, Chimney Swift, Ruby-throated Hummingbird, Eastern Phoebe, Kinglets, Purple Finch, American Goldfinch.

Winter Your best bet is at the Lake and other bodies of open water for overwintering waterfowl, including Wood, Ring-necked, and Ruddy Ducks; Hooded, Common, and Red-breasted Mergansers. A number of sparrows spend at least part of the winter here, too, including American Tree, Fox, Song, Swamp, and White-throated as well as Dark-eyed Junco.

HABITAT

Woodland, edge, lake, marsh, freshwater.

BEST TIME TO GO

Spring and fall during migration; lots of nesting birds in summer and nice waterfowl in the winter.

Orange-crowned Warbler

HOW TO GET TO PROSPECT PARK

By Subway The ❷ and ❸ stop at Grand Army Plaza, the ideal entrance. The park is also served by the Ⓓ Ⓕ Ⓖ Ⓠ and ❹ at various entrances (see map) and by a number of public bus lines.

By Car Prospect Park is less than a thirty-minute drive (in light traffic) from Manhattan. Parking is available along the streets, at the Brooklyn Museum, and in a lot off East Drive in the southeastern part of the park.

INFORMATION

Further information is at http://www.prospectpark.org

During the spring and fall, the Brooklyn Bird Club conducts walks for members and guests. Their guides provide an excellent overview of the best areas and are welcoming and knowledgeable. See http://www.brooklyn birdclub.org

The Brooklyn Botanic Garden, adjacent to Prospect Park, is an absolute jewel, especially in spring. The birdiest area is the Native Flora section, which, coincidentally, is nearest Prospect Park. You can even learn what's in bloom (at Bbg.org) in order to plan your trip. Also nearby is the renowned Brooklyn Museum, as well as the historic Green-Wood Cemetery, another great birding spot. You can find more information on Green-Wood and Brooklyn Botanic Garden in this chapter.

Scope Not necessary.

Photos Easy.

Brooklyn Botanic Garden

No trip to Brooklyn is complete without a visit to the Botanic Garden. Not only are these delightful gardens good for the soul—they are also attractive to birds. Transformed from land that was an ash dump in the late 1800s, the garden's fifty-two acres are now one of the most beautiful parks in New York City. Located near Prospect Park and the Brooklyn Museum, Brooklyn Botanic is easy to get to and enjoy. If you are not a member, you may have to pay a fee, but it's well worth the few dollars to leave the city behind and immerse yourself in the charms of this stunning little park and its variety of carefully maintained habitats. This is a wonderful place to visit for both its beauty and variety of birds in certain areas. Security throughout makes it a safe place to bird alone.

VIEWING SPOTS

Get your ticket and map at the Eastern Parkway entrance gate. If you are short on time, walk through the *Osborne Garden*, stay on the right main pathway at the end of this section and head straight to the *Native Flora Garden*. During migration this is an interesting and busy area for warblers and other migratory songbirds, sparrows, and wrens. Year-round, Red-bellied and Downy Woodpeckers are seen. It's a small area, but take advantage of its proximity to Prospect Park across the street and this section's well-cared-for trails to note how popular the native plantings are with birds. Not surprisingly, it's probably the birdiest and most reliable spot in the garden.

If you have more time, treat yourself and explore the rest of the BBG. Exiting the Native Flora Garden, investigate the *Cranford Rose Garden* (rabbits may be seen near the building at the north end), and the popular *Cherry Esplanade*. Check the grass and low shrubs of these areas for ground feeders like Northern Flickers, White-throated and Chipping Sparrows, and, in winter, Dark-eyed Juncos. The fruit trees northeast of the Cherry Esplanade may have migrants in season, and be on the watch for fallouts, which happen from time to time.

Continue on to the *Japanese Hill-and-Pond Garden*. This Zen area is a highlight of the garden, and there is often a good amount of bird activity in the trees around it, as well as in the pond. Several species of ducks ply the waters here, including Wood Ducks and Pied-billed Grebes. A variety of birds may nest in the pagoda and other structures.

The buildings housing the *Library* and *Palm House* are fronted with low hedges and shrubbery, which are worth a look. The *Lily Pool* sometimes has waterbirds, and Great Blue Herons find the goldfish irresistible. In spring, skulking wrens and Common Yellowthroat are found there. Make sure you keep your binos and camera handy when having lunch or tea in the *outdoor café*, as you never know what birds might be prowling around the edges.

If you have time for a further walk, take in the *Plant Family Collections* at the garden's far south end; the *Terminal Pond* may also be productive.

Overhead, keep a look out for Red-tailed Hawks, Peregrine Falcons, and Barn and Tree Swallows. Look on the ground for Northern Flicker, Mourning Doves, Eastern Towhee, Chipping Sparrow, and other ground feeders.

KEY SPECIES BY SEASON

Spring Expect warblers, vireos, and a variety of other songbirds during migration. Common Yellowthroat, Northern Parula, Redstart, Magnolia, Black-and-white, Yellow, Black-throated Blue, Louisiana Waterthrush, and Northern Waterthrush are some of the more than two dozen species of

Female American Redstart

warbler that stop here on migration; Eastern Towhee, Great Blue Heron, and Peregrine Falcon can be seen spring through fall.

Summer Baltimore Oriole, Chimney Swift, Eastern Kingbird, and a variety of nesting songbirds.

Fall Hermit Thrush, Brown Creeper, Golden and Ruby-crowned Kinglets, and the same wealth of migrating warblers that come through in springtime.

Winter Yellow-bellied Sapsucker, Golden-crowned Kinglet, Dark-eyed Junco, Carolina Wren, White-throated Sparrow, and waterfowl. It's a good time to see resident species such as Black-capped Chickadee, Tufted Titmouse, Song Sparrow, Red-breasted Nuthatch. Red-bellied and Downy Woodpeckers and Red-tailed Hawk are seen year-round.

HABITAT

Wooded areas, native plants, freshwater, edges.

BEST TIME TO GO

April through November will give you the best chance of seeing the most birds. Migration brings songbirds through the gardens, as well as the occasional hummingbird visiting the Herb Garden. There are nesting birds throughout. A number of berry-producing shrubs and trees attract migrants in the fall, including Cedar Waxwings and White-breasted Nuthatches. Look for hawks and woodpeckers in winter.

HOW TO GET TO THE BROOKLYN BOTANIC GARDEN

There are two entrances—Washington Avenue and Eastern Parkway.

By Subway **2** or **3** to the Eastern Parkway–Brooklyn Museum station; **B** **Q** or **S** to the Prospect Park station; **4** or **5** to the Franklin Avenue station.

By Bus Those that come closest to the two entrances are the B48 to Eastern Parkway and B45 to Washington Avenue. The following stops are about a ten-minute walk from one of the garden's entrances: the B49 to Eastern Parkway, B43 to Washington Avenue, B41 or B69 to Brooklyn Public Library, B65 to Classon Avenue, and B16 to Empire Boulevard.

By Train Long Island Rail Road to Flatbush Avenue/Atlantic Avenue station. Connect with **2** **3** **4** or **5** or B41.

By Car If you drive, there is parking at 900 Washington Avenue with hourly charges.

150 Eastern Parkway, 990 Washington Avenue, Brooklyn, NY 11225
718-623-7200
http://www.bbg.org

The garden charges an entrance fee for nonmembers, but there are free days, too. The website has up-to-date information and provides a map.

Restrooms are found at the Washington Avenue entrance, as well as in the library building and the Steinhardt Conservatory, where you'll find a café that serves food year-round. The shop is terrific.

The BBG is a great spot for a birding trip with friends who might not be interested in birding. They might also be lured by the adjacent Brooklyn Museum, and Green-Wood Cemetery is just a taxi ride away—see its full description in this chapter. The Botanic Garden borders Prospect Park along Flatbush Avenue and makes a productive, full-day trip when you visit both.

Scope You won't need it.

Photos Easy.

Green-Wood Cemetery

Green-Wood is one of the first rural cemeteries in the United States and the site of the Battle of Long Island in 1776. It is now a National Historic Landmark with tranquil grounds, jaw-dropping views, and elaborate graves and mausoleums. It was founded in 1838, and over a half million people are buried here, including celebrities like conductor Leonard Bernstein, founder of the *New York Tribune* Horace Greeley, cabinetmaker Duncan Phyfe, and the infamous "Boss" Tweed. There are also infamous birds, such as the flock of Monk Parakeets that have taken up residence in the gothic spires of the gatehouse. Green-Wood is a productive and secluded spot to see birds on its 478 acres of wooded areas, edges, and ponds. Year-round, this makes for a peaceful and unusual birding experience, and one you can easily do on your own.

VIEWING SPOTS

To focus on the well-known birding hot spots within the cemetery, enter through the imposing *Main Gate* at 25th Street and Fifth Avenue. You may be greeted by the sounds, if not the sight, of the exotic Monk Parakeet colony

Ⓓ Ⓝ Ⓡ   4th Ave.   Ⓓ Ⓝ Ⓡ

35th St.   5th Ave.   35th St.   24th St.   23rd St.   22nd St.   21st St.   20th St.

● Entrance   B63 ■   ● Main Gate

Monk Parakeets

Gothic Spires ●

36th St.   6th Ave.

SYLVAN LAKE   Lake Ave.   VALLEY WATER   Chapel

Landscape Ave.   7th Ave.

Lake Ave.   PIERREPONT HILL   Sycamore Ave.

Oak Ave.   Central Ave.

37th St.

Landscape Ave.

Hillock Ave.   Vista Ave.   Prospect Park West

Orchard Ave.   Pine Ave.   Entrance ●

Crescent Ave.   Forest Ave.

Ninth Ave.   DELL WATER   Dale Ave.   Locust Ave.   Woodland Ave.   OLD BURIAL GROUND

Ⓓ   Vale Ave.   CRESCENT WATER   Vernal Ave.   Central Ave.

Southwood Ave.   Peter Cooper Circle

Grove Ave.   OCEAN HILL

Cypress Ave.   Ocean Ave.

Vine Ave.   Entrance ●   Ft. Hamilton Pkwy.

Minna St.   McDonald Ave.   E. 2nd St.   E. 3rd St.

37th St.   36th St.   12th Ave.   Chester Ave.   Dahill Rd.

Tehama St.

Clara St.

N ↗

GREEN-WOOD CEMETERY

Eastern Kingbird in Green-Wood Cemetery

that nests there. Pick up a map here, as the cemetery is sprawling, full of meandering paths, and offers many opportunities to get lost.

Inside the gate make either a right or a left toward Central Avenue, then stop *behind the chapel*, which is a great spot for birds.

Scan the huge *Copper Beeches* and other mature trees on the hill not far from the entrance. Not only does this location offer great views of the cemetery, but the magnificent trees themselves attract birds. During migration, warblers can be seen literally hanging off the branches. Another warbler magnet is *Cypress Avenue*.

Don't miss visiting the freshwater ponds. *Sylvan* and *Valley Water* are larger, while *Crescent Water* is a smaller, more intimate location where you may find a Black-crowned Night-Heron lurking, or the odd migrating sandpiper.

During the various stages of migration, over thirty species of warbler can be found, as well as waders, shorebirds, hawks, and ducks. Year-round residents include Great Blue Heron, Belted Kingfisher, Red-bellied and Downy Woodpeckers, Northern Flicker, Red-tailed Hawk, Black-capped Chickadee, and Red and White-breasted Nuthatches.

Avoid a lot of extra walking by sticking to the area from the main gate to the four bodies of water and the paths indicated on our map. In general, the closer you get to the perimeter of the cemetery, the less interesting the birding becomes.

KEY SPECIES BY SEASON

Spring Warblers galore, including Black-throated Blue, Yellow, Nashville, Common Yellowthroat, Northern Parula, Chestnut-sided, Blackpoll, Oven-bird; Blue-gray Gnatcatcher, Eastern Bluebird, Swainson's Thrush.

Summer Eastern Kingbird, Baltimore Oriole, Red-winged Blackbird, Black-crowned Night-Heron.

Fall American Kestrel, Cooper's and Sharp-shinned Hawks, songbirds, Eastern Phoebe.

Winter American Wigeon, Northern Shoveler, Snow Goose, Hooded Merganser, Dark-eyed Junco, White-throated Sparrow, Great Horned Owl. Year-round, Great Blue Heron, Belted Kingfisher, some hawks and woodpeckers, along with more common winter residents listed above.

HABITAT

Woodlands, edges, and freshwater.

BEST TIME TO GO

April through October will have the most birds, but this tranquil spot is a great change-of-pace location year-round.

HOW TO GET TO GREEN-WOOD CEMETERY

By Subway Take the ⓡ and exit at the 25th Street station. Walk east one block to Green-Wood at Fifth Avenue and 25th Street.

By Car If you are using GPS, use the address "25th Street and Fifth Avenue, Brooklyn, NY." Do NOT use "500 25th Street."

Free parking is available within Green-Wood. Please park all vehicles on the right-hand side of the road. Parking on the grass is prohibited.

INFORMATION

The main entrance is at the gothic arches at Fifth Avenue and 25th Street. The other entrances, at Fourth Avenue and 35th Street, at Fort Hamilton Parkway, and on Prospect Park West, may have different opening times than the main gate. During inclement weather, some of these entrances may be closed.

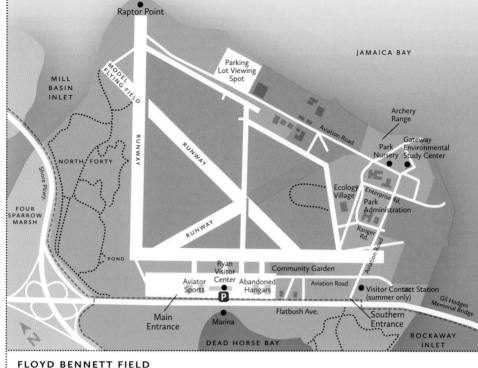

Raptor Point

JAMAICA BAY

Parking
Lot Viewing
Spot

MILL
BASIN
INLET

MODEL
FLYING FIELD

Archery
Range

Aviation Road

Park
Nursery

Gateway
Environmental
Study Center

RUNWAY

RUNWAY

NORTH FORTY

Shore Pkwy.

Ecology
Village

Enterprise Rd.

Park
Administration

FOUR
SPARROW
MARSH

Ranger
Rd.

RUNWAY

Aviation Road

POND

Ryan
Visitor
Center

Community Garden

Aviator
Sports

Abandoned
Hangars

Aviation Road

P

Visitor Contact Station
(summer only)

Gil Hodges
Memorial Bridge

Flatbush Ave.

Main
Entrance

Marina

Southern
Entrance

ROCKAWAY
INLET

DEAD HORSE BAY

FLOYD BENNETT FIELD

For more information call 718-768-7300 or e-mail info@green-wood.com

http://www.green-wood.com will provide a map
http://www.nycaudubon.org
http://www.brooklynbirdclub.org

*Please note that Green-Wood is an active and working cemetery. Be respectful of
funeral services and those visiting loved ones.*

Scope Not necessary.

Photos Easy.

Floyd Bennett Field

Most airfields don't encourage birds, but New York City's first municipal
airport, founded in 1931 and now on the National Register of Historic Places,
is a birding hot spot. These days the only flying objects with motors are the
NYPD helicopters—the rest of the traffic is purely avian. With 140 acres of
its 1,400 total acreage as grassland, you can expect to see raptors like the
resident Marine Parkway Bridge Peregrines hunting over the fields. There

BROOKLYN

Snowy Owl at Floyd Bennett Field during irruption

are also grassland birds—lots that migrate through—and over two dozen species that nest here in summer. It's also great for winter waterfowl, and during an irruption, Snowy Owls can be seen at close range. Because Floyd Bennett is so huge, you must have a car or bicycle to thoroughly explore it.

VIEWING SPOTS

Floyd Bennett has two entrances, both off Flatbush Avenue. The northern entrance, which you would use for Aviator Sports, will provide easy access to the Ryan Visitor Center. There you can use the restrooms, get information from rangers on duty, or browse the gift shop.

A drive around the *grasslands* using the roads or old runways will be productive in all seasons. In winter especially, make your way out to *Raptor Point* and to the nearby large parking lot that faces Jamaica Bay. Not only will you find waterfowl, but possibly Northern Harrier, and in the parking lot, Snow Bunting. General birding etiquette is to remain at the bulkhead so as not to disturb the birds. From the shore, you can also see a couple of islands that are inaccessible and offer good breeding habitat for waders and other colonial nesters.

In winter, Horned Larks may be found on the short lawn by the *visitor center* and *abandoned hangars*. In spring, summer, and fall, a walk into the *North Forty* is worthwhile if you visit *the pond*. Two blinds allow up-close viewing, but you'll ruin your chances of seeing much if you make noise as you approach. For birders, there's no advantage to continue hiking in the North Forty after you've seen the pond.

On the other side of the park, a *stand of pitch pines* near the southern entrance is reputed to be good for owls in winter. Something interesting might be lurking at the *Community Gardens* as well. Other areas like the Ecology Village or the ruined buildings by the Archery Range could have some nice birds, but not generally, so give them lower priority if you're pressed for time.

Keep in mind that large parts of Floyd Bennett are used by the Department of Sanitation and the New York City Police Department and are off-limits to the public.

KEY SPECIES BY SEASON

Spring Wood Duck, Green-winged Teal (also in fall and winter); American Woodcock; Fish Crows stay through the breeding season; some warblers.

Summer Wading birds nest in the area, and so do herons, egrets, Glossy Ibis, Osprey, American Oystercatcher, Willet, Laughing Gull, terns, Black Skimmer, Willow Flycatcher, White-eyed and Warbling Vireos, Tree and Barn Swallows.

Fall Migrating songbirds, especially sparrows, and hawks. It's a good location for shorebirds beginning in late July, including the chance to see something unusual besides Least and Semipalmated Sandpipers. Dunlin overwinter; American Kestrels, Merlin, and Peregrine Falcons arrive and stay the winter. The best place in the five boroughs to see Blue Grosbeak, Dickcissel and Bobolink—but none are believed to breed here.

Winter Waterfowl in the pond and open water including Bufflehead, mergansers, primarily Greater Scaup rather than Lesser, American Wigeon, Common and Red-throated Loons; Snow Geese are occasionally seen; Horned Grebe, Great Cormorant; a perfect habitat for Snowy Owls; possible Northern Harrier; Horned Larks, Snow Buntings; overwintering sparrows. Uncommonly, American Pipit.

HABITAT

Open grasslands, edges, marsh, saltwater, some forest.

A deservedly popular birding destination in winter, great in fall, good in spring.

HOW TO GET TO FLOYD BENNETT FIELD

BY BUS Take the Q35 southbound on Flatbush Avenue and get out at the Floyd Bennett Field stop by the Gateway Marina.

BY CAR The official address for Floyd Bennett is 50 Aviator Road. We recommend that you set your GPS for 3159 Flatbush Avenue, Brooklyn, NY 11234, which is the entrance to Aviator Sports and the closest entrance to the Ryan Visitor Center.

INFORMATION

New York Harbor Parks website: http://www.nyharborparks.org
National Park Service website: http://www.nps.gov
NYC Parks website (with detailed driving directions):
    http://www.nycgovparks.org
Brooklyn Bird Club: http://www.brooklynbirdclub.org
NYC Audubon: http://www.nycaudubon.org
Birdwatching Daily: http://www.birdwatchingdaily.com

OTHER THAN BIRDING

This is the only park in New York City that allows overnight camping. You'll also find model-plane flying, kayaking, fishing. The National Park Service conducts a number of public programs in gardening, stargazing, and biking, among others. Ample parking. The sports center has ice-skating and other activities and sells snack food and drinks.

Dead Horse Bay and Dead Horse Point

Directly across Flatbush Avenue as you leave Floyd Bennett Field is a very good stop to make any time of year. We mention it in our Coastal Brooklyn Winter Waterfowl Viewing section since winter is the best time to visit Dead Horse Bay and Dead Horse Point. But it has its attractions in other seasons as well. Park in the lot near the southern entrance to Floyd Bennett at Aviation Road and walk across Flatbush Avenue, or get off the Q35 bus near the tollbooth to the Marine Parkway Gil Hodges Memorial Bridge. Enter this area at the intersection of Flatbush and Aviation. After a few yards, take the path to the right and make your way to the beach. Then follow the beach to the left and around the point toward the bridge. To use this route you must arrive at low tide, which is the best time to see shorebirds here anyway. In winter, look for a wide variety of waterfowl from the beach. Spring attracts migrating shorebirds. In summer, look for waders, American Oystercatcher, Least and Common Terns, and gulls of several species; also some songbirds, including Eastern Towhee. In fall, there are a variety of migrating sparrows and warblers.

COASTAL BROOKLYN WINTER WATERFOWL VIEWING

Coastal Brooklyn provides an excellent opportunity to see some of the diverse and unexpected urban places giving refuge to winter waterfowl. This area is surprising in a number of ways. Rarities turn up, and who would think you could see thousands, much less tens of thousands, of any one species from one spot in New York City? Because Brooklyn's coast offers so many opportunities for viewing waterfowl, we have compiled the following guide to the most popular ones. Time your visit from November to March—the highest concentration of birds typically occurs after the first of the year. None of these locations are going to transport you to the wilderness, but they are interesting for the variety and sometimes astonishing abundance of birds.

A full tour of all the stops would start at Brooklyn Bridge Park and take you all around the coast of Brooklyn to Jamaica Bay. However, the purpose of this section is not to create a full day (or longer) tour. If you are looking for ducks, we offer a list of spots that are generally productive in season. Be prepared for an ample number of gulls—in some locations there will be hundreds of them. You might also see an interesting pelagic or shorebird.

Since some of these waterfront hot spots are actually worth a visit in other seasons, they are described separately in this chapter. Some—especially those from Gerritsen to Spring Creek—can easily be combined with a trip to Jamaica Bay or Floyd Bennett Field (both described separately).

As a safety precaution, we highly recommend you visit the majority of these locations with a buddy. Many are not well-visited, and most do not have security.

A scope is often recommended.

Coastal Brooklyn covers a large expanse, but it divides itself into three sections. The X factor will be your relationship to Interstate Highway 278 and the Belt Parkway. For stops between Brooklyn Bridge Park and Owls Head/American Vets Pier, stay off 278. Beyond that, get on the Belt Parkway to gain access to the rest. If you are an avid biker and immune to cold weather, you can leave the car at home.

At most of these stops, you can reasonably expect to see a variety of waterfowl, ranging from the odd accidental to huge rafts of Brant. The most common winter waterfowl include Canada Geese, Brant, Gadwall, Red-breasted and Hooded Mergansers, Greater Scaup, Horned Grebe, Bufflehead, Loons, American Wigeon, Common Goldeneye; larger open water may attract Long-tailed Ducks and larger rafts of birds. Each location will have a different variety and abundance based on its own peculiarities. Ring-billed, Herring, and Greater Black-backed Gulls abound, but Bonaparte's, Glaucous, and even Iceland Gulls are sometimes seen. Some locations offer the possibility of Northern Gannet, scoters, and the occasional pelagic.

Especially productive places are shown with an asterisk.

Brooklyn Bridge Park*

Expect the normal array of harbor waterfowl, including Gadwall, Brant, Bufflehead, Mallard, and Red-breasted Merganser—see our expanded description of Brooklyn Bridge Park in Other Places to Find Birds in Brooklyn.

Perfectly civilized, this is one location where you can find cafés and restrooms, and it is easily accessible from Manhattan by the East River Ferry; nearby subway stops are ❷ ❸ Clark Street; ❹ ❻ High Street; ❺ York Street. The park encompasses six piers. If you drive, you might be lucky enough to find metered parking along Furman Street near Piers 2 and 3.

http://www.brooklynbridgepark.org

Bush Terminal Piers Park*

This is another amazing transformation of the Brooklyn waterfront into a lovely modern park. Access the entrance at First Avenue and 43rd Street, where some parking is available. Bring your scope. You'll get views of the sheltering waters between rotting piers, as well as the open waters of Bay Ridge Channel and the Upper New York Bay. A bird-attracting, man-made double impoundment allows you to get close to your quarry. This can be a productive winter stop for gulls and waterfowl, including American and Eurasian Wigeon, nice numbers of scaup, and many others. Restrooms are open year-round. Information about birding here in other seasons can be found in Other Places to Find Birds in Brooklyn.

http://www.nycgovparks.org

Brooklyn Army Terminal Pier 4

Access this commuter dock via First Avenue and 58th Street by driving or walking down 58th Street toward the water and following the signs for the ferry to Manhattan. If you go on the weekend it will be virtually deserted and you'll be able to drive your car out onto the large concrete pier and commune with the gulls. Park anywhere you want and look at the waters on either side for waterfowl. No restroom facilities.

Owls Head Park and American Veterans Memorial Pier

If you park along 68th Street, you'll have ready access to the American Veterans Memorial Pier. Interesting waterfowl are sometimes found here, along with large numbers of Bufflehead—from a couple of dozen to over one hundred—and on fortuitous occasions, Bonaparte's Gull. Please also note that while the nearby water treatment plant looks like it might have some great shoreline birding, it's not open to the public. Don't forget the scope. See Other Places to Find Birds in Brooklyn for what's happening in Owls Head in other seasons. Restrooms are open year-round.

To reach the next stop and all of the following, you must use the Belt Parkway. Drive toward the water on 68th Street and make a right (although you'll eventually be going left) on Shore Road. This takes you around the north side of Owls Head Park to Second Avenue, where you make a left and another quick left to the Belt.

http://www.nycgovparks.org

Gravesend Bay*

Just north of the Verrazano-Narrows Bridge is the first of a series of coastal pull-overs from the Belt Parkway that can lead to good views of waterfowl from this long strip of a park. The first stop is known as the Narrows, just before the bridge footings. Walk under the bridge and around the corner to a little ruin of a pier that has a small beach. Check the waters along here, especially around the bridge footings, where ducks may congregate. When you are done here, drive south on the parkway and, over the next few miles, look for pull-overs, but be very careful getting back on the parkway. In fact, when leaving the Narrows, make sure to merge into the second lane—not the closest one. That is the only way to avoid the ramp that takes you off the Belt and onto the Fort Hamilton Parkway or Verrazano-Narrows Bridge, which would make your next stop Staten Island.

The next few miles, including the *Middle Parking Lot* (just after you pass under the bridge), *Dyker Beach*, and the *South Parking Lot*, may offer some interesting birds. Scaup and Horned Grebe may be found, along with other waterfowl or unusual gulls.

Calvert Vaux Park*

Formerly Drier-Offerman Park, this park comes to us courtesy of the Verrazano Bridge, for it is from the sand and excavated rock dredged up in the building of the bridge that this park was created. Bypass the first parking lot (unless you need a restroom stop) and drive closer to the water. Once on foot, head for the bay. You may be rewarded with sightings of waterfowl—and sometimes lots of them, including Red-breasted Mergansers, Bufflehead, Horned Grebe, Common and Red-throated Loon, and Common Goldeneye during the winter. Rafts of over one thousand scaup have been reported. Check for Killdeer, Horned Lark, and interesting sparrows, some of which may also be lurking on the unused baseball fields to the north. For information on Calvert Vaux Park in other seasons see Other Places to Find Birds in Brooklyn.

Be aware that Calvert Vaux Park can really accumulate trash along the shore. Not many people use the lonely area near the water, so make sure you take along a friend.

http://www.nycgovparks.org

Coney Island Creek*

Coney Island Creek is definitely worth the stop if you're prepared for a heavily polluted former industrial water dump. Park on the street near West 23rd

Street and look out into the waters of Coney Island Creek for an often amazing collection of waterfowl.

An illuminating photo-essay by Nathan Kensinger, "Coney Island's Untamed Creek, Caught between Past and Future," can be found at http://ny.curbed.com.

Coney Island Creek Park

Park along Bayview Avenue between West 35th and West 37th Streets and climb over the dunes to the wide beach on the other side. It's much more scenic and safe-seeming than our two previous stops, but the birding is not as good. It is a reasonably OK spot to look for Snow Buntings and migrants in spring and fall if you happen to be in the area.

http://www.nycgovparks.org

Coney Island Pier

Rebuilt after Hurricane Sandy, this one-thousand-foot pier takes you out into Coney Island Channel. Here you may find Long-tailed Ducks in decent numbers, scaup sometimes in rafts of over one hundred, Brant, and Common Loon. Northern Gannet have been reported as well. Accessible from Surf Avenue via the Boardwalk at West 15th or West 21st Street.

http://www.nycgovparks.org

Plumb Beach*

Plumb Beach (sometimes in reports as "Plum Beach") is well worth the visit. Scaup can be found—sometimes in the tens of thousands; Long-tailed Ducks, Bufflehead, large rafts of Brant, and Red-breasted Merganser. To access Plumb Beach, head east on the Belt Parkway. There is no sign for it until you are practically there, so after Exit 9, look for a blue sign directing you to a rest area, and follow that to the parking lot. The kiosk and its restrooms will be closed if you're coming in winter, but there may be portable toilets available. Once on foot, follow the beach eastward to Point Breeze at the mouth of Gerritsen Creek. In addition to waterfowl, you might catch a shorebird or two. Find information on birding here in other seasons in Other Places to Find Birds in Brooklyn.

Glossy Ibis

Osprey on nest

At this point, you're practically at the doorstep of *Floyd Bennett Field*—which is described earlier in this chapter. If Floyd Bennett is your destination, get back on the Belt/Shore Parkway and take Exit 11S. But if you want to make one last stop, use Exit 11N, going north on Flatbush Avenue and left on Avenue U near the Kings Plaza Shopping Mall for the *Salt Marsh Nature Center at Marine Park*.

Plumb Beach: http://www.nycgovparks.org

Salt Marsh Nature Center at Marine Park

Marine Park is a 530-acre saltwater marsh surrounding Gerritsen Creek. Most of it is covered in high phragmites, so while the grasses may be teeming with birds and wildlife, viewing is difficult. If you go, we recommend a visit to the Salt Marsh Nature Center (restrooms are found here when the building is open) and take a short stroll out to see Gerritsen Creek. Waterfowl, wading birds, shorebirds, marsh birds, and fall sparrows are all seen here. Because of its proximity to Jamaica Bay, there are terns and Osprey in season. If you want to hike, continue on the trails, but the birding is not particularly satisfying. Plans are under way to redo the trail system, so in future this may become a recommended stop.

Park at the Salt Marsh Nature Center on Avenue U and 33rd Street, Brooklyn, NY 11234.

Or you can take the Ⓓ Ⓕ Ⓝ or Ⓠ to Avenue U, then the B3 along Avenue U to the Nature Center and park entrance.

718-421-2021
http://www.nycgovparks.org
http://www.saltmarshalliance.org

Dead Horse Bay and Dead Horse Point

Dead Horse Bay and Dead Horse Point is an excellent winter waterfowl spot close enough to Floyd Bennett Field that you can park at Floyd Bennett's southern entrance on Aviation Boulevard and walk across Flatbush Avenue to the beach. Plan to make this trip at low tide. There can be some unusual waterfowl from time to time, and this site has a reputation for attracting birds that are off course or disoriented, so it's worth a visit to see what has made its way here. More information on visiting this site in other seasons

and where to view birds can be found earlier in this chapter under Floyd
Bennett Field.

Canarsie Pier*

Canarsie Pier is a good birding spot favored by fishermen and used by locals
as a hangout. There are Bufflehead, Horned Grebe, and a number of other
birds, as well as gulls. It is also a quick and easy in and out by car. If you're
lucky and have your scope you'll see a nice bird or even a seal hauled out on
the little island across the channel in Jamaica Bay.

    http://www.nyharborparks.org

Canarsie Park

Canarsie Park is a public park and playground through which you walk to get
to Paerdegat Basin, where there may be ducks in the harbor, and it's worth
a scan to the left for waterfowl farther afield. It is possible to walk on the
beach to see ducks that may be a bit farther away. Bufflehead can be found
in decent numbers, and there can be abundant Brant.

    http://www.nycgovparks.org

Scaup flying in Hendrix Creek

Fresh Creek Park

Fresh Creek Park contains a forty-two-acre Forever Wild site and wetlands preserve. Stop first at 108th Street between Flatlands 3rd and 4th Streets, where there is an opportunity to get a view over much of the marsh. Hooded Merganser, Bufflehead, and American Wigeon like this area. Make another stop on the opposite side of the creek by taking 108th to Flatlands Avenue, making a right and then the next right on Louisiana Avenue. The official entrance is a few blocks in, and after walking straight out you'll find an overlook onto the salt marshes you just passed. Bring that scope!

http://www.nycgovparks.org

Hendrix Creek* and Betts Creek

Hendrix Creek is a ribbon park wedged between a shopping mall and a wastewater treatment plant. Park in the JC Penney lot and use the crosswalk to head toward the plant. Find a path to the water and check out the waterfowl—Brant, Bufflehead, Ruddy Ducks, wigeons, and shorebirds like yellowlegs can be found here. Look left, and if it appears the birds are farther down the water, you can take the pedestrian paths along the "creek."

Looming over the scene are two former landfills, still off-limits to visitors. From the JC Penney parking lot, make a left onto Gateway, which becomes Seaview. At its dead end is Fountain Avenue. Park here to look into Betts Creek. To your right, going south on Fountain, is the infamous Fountain Avenue Landfill—a toxic waste and disposal area allegedly used by the mob. It's been capped and, like its sister, the Pennsylvania Avenue Landfill, will someday be open to the public. Like many of these sites, the afterlife of a dump is a prairie with lots of wildlife. At the time of writing, both are still off-limits, but it looks like they will eventually offer a wonderful habitat for grassland birds. Keep checking to see the current status. For more background see the article by Kenneth Chang at http://www.nytimes.com, "A Wooded Prairie Springs from a Site Once Piled High with Garbage."

Spring Creek Park

Spring Creek Park is technically in Queens, but since it is potentially a good site to see waterfowl and just on the Brooklyn border, we have included it here.

It may have a name that conjures up images of a pretty setting, but at present it is a trash-laden and abused wild area. Often rafts of ducks and geese spend the winter; shorebirds poke around in the muck and debris in late summer and fall. Park along 161st Avenue where it dead-ends into 78th Street. Huge swaths of phragmites with winding paths cut through them stand between you and the water. It's not considered a safe location, so having a friend or two is absolutely essential here. A GPS locator will help you track your way through the maze of weeds to the water, where you will pick your way through broken glass and garbage of all sorts on the beach. Ducks bob in the water alongside discarded plastic bags and bottles. It's a sad example of what birds, often desperate for habitat, have to contend with in a heavily populated area. A scope would be a big help here, as the birds can be far away. This location is good for a quick walk out and back if you keep that GPS tracking on.

http://www.nycgovparks.org

OTHER PLACES TO FIND BIRDS IN BROOKLYN

There are a number of small parks in Brooklyn that make for productive birding year-round. Some are already noted in Coastal Brooklyn Winter Waterfowl Viewing. Here's the rundown in other seasons. The most productive are marked with an asterisk.

Brooklyn Bridge Park*

These eighty-five acres of Brooklyn waterfront are undergoing a striking transformation. With over one mile of coastline, killer views of Manhattan, and lots of great things to do, it's a nice spot to take a stroll and do some birding as well. Salt marshes, meadows, and native woodland gardens attract birds to this location. It's not a huge hot spot yet, although with careful development, the bird-friendly habitat may continue to improve. In short, it's a safe and pleasant place to look for birds in a historic and popular area. You will find cafés and restrooms, and it is easily accessible from Manhattan by the East River Ferry. Nearby subway stops are ❷ ❸ Clark Street; Ⓐ Ⓒ High Street; Ⓕ York Street. It's a long shot, but you may find metered parking along Furman Street near Piers 2 and 3. Also good in winter—see Coastal Brooklyn Winter Waterfowl Viewing.

Bush Terminal Piers Park*

This reclaimed landfill is so nestled in an industrial neighborhood that it's hard to believe it even exists. Access the entrance at First Avenue and 43rd Street, where some parking is available, and walk along the waterfront esplanade past the tide ponds and restored wetlands. Restrooms are open year-round. It's also good in winter—see Coastal Brooklyn Winter Waterfowl Viewing. If you are taking public transportation, you will need to walk about ten minutes from the 45th Street stop on the **N** or **R**.

Owls Head Park and American Veterans Memorial Pier

Great views of ships entering the harbor are to be found at this popular park. We have highlighted it for winter—see our section on Coastal Brooklyn Winter Waterfowl Viewing—since it provides access to the American Veterans Memorial Pier. Owls Head is a nice, short outing spring and fall as migrant songbirds pass through. Interesting fall sparrows find their way here as well. Going by car is best, with parking along 68th Street, or it's about a ten-minute walk from the Bay Ridge stop on the **R**.

Calvert Vaux Park*

Calvert Vaux Park—formerly Drier-Offerman Park—is a good spot for birding in all seasons, but make sure that you do not go here alone. Look for Great Blue Heron and other waders, terns, and shorebirds spring through fall; in summer, Spotted Sandpiper, Green Heron, Killdeer, as well as Osprey, Common Tern, Monk Parakeet, Willow Flycatcher, and a variety of other breeding songbirds. Terrific in winter—see Coastal Brooklyn Winter Waterfowl Viewing. Going by car is best as it's about a fifteen-minute walk from the Bay 50th Street stop on the **D**. You can cross Shore Parkway via the overpass on 27th Avenue.

Plumb Beach*

Sometimes written about as "Plum Beach." Low tide is best to see shorebirds from this spit of land with mudflats and a salt lagoon. Check for waders spring through fall, including Glossy Ibis; shorebirds like American Oystercatcher,

Willet, Spotted Sandpiper, "peeps," and a variety of others; Black-bellied Plovers and Black Skimmers, especially in fall; Least Terns in summer; some songbirds spring through fall—often seen from the low shrubs near the parking lot. Also good in winter—see Coastal Brooklyn Winter Waterfowl Viewing. Please note that it has a reputation as a popular "cruising" spot as well. Access by car traveling east only from the Belt Parkway: look for a blue sign that directs you to the rest area after Exit 9 and get off at the rest area.

Four Sparrow Marsh

Four Sparrow Marsh has been a wild and undisturbed marshland supporting nesting Swamp, Saltmarsh, Song, and Savannah Sparrows, along with over one hundred other migrating and overwintering bird species. Even though it was once designated Forever Wild, that status may not protect it from development. As of this writing, its entrance near the Toys R Us is closed, and while the marsh still may support waders, sparrows, songbirds, and winter waterfowl, we cannot recommend the site for safety reasons. If you want to visit this little marsh, view it from the sidewalk along Flatbush Avenue or the Shore Parkway, which is less than ideal. All birders hope that this once-thriving wetland is restored and maintained as a wild habitat.

Established in 1683 as one of the original twelve New York counties, Queens is the easternmost of the five boroughs, bounded by Brooklyn on the west and Long Island's Nassau County on the east. If you have used either of the two New York City airports, you have already been to Queens, as both LaGuardia and JFK are located here. Named for Catherine of Braganza, the wife of King Charles II, it is only fitting that a borough whose population is nearly evenly divided between American-born residents and immigrants is named after an English queen born in Portugal. Her spirit presides over what is now the second-most-populous county in New York State and one of the most ethnically diverse urban areas anywhere. It may therefore come as a surprise that Queens also has diverse natural habitats and great places to see birds.

The *Jamaica Bay Wildlife Refuge* is the long-reigning king of birding sites in Queens. Despite the breach in the West Pond left by Hurricane Sandy that changed the pond from a freshwater lake to a saltwater lagoon, the refuge still has amazing shorebirds. The bird life is different now at this Important Bird Area, but it is still at the top of the list for Queens birding and should not be missed. The general consensus is that for this park to return to its status as a legendary birding hot spot attracting a huge diversity of birds, the breach needs to be repaired. Several environmental and community groups are working to do just that.

Meanwhile, there is a lot more to Queens birding than just Jamaica Bay. Birds love the mature woods at Olmsted-designed *Forest Park*; *Alley Pond Park* offers wetlands and forest wedged in between several busy highways; wild beaches host endangered nesting birds at *Breezy Point*. Nearby, *Jacob Riis Park* and the retired coastal army base at *Fort Tilden* combine great beach, woods, and scrub habitat with an excellent hawk watch. For a more intimate experience, try the *Queens Botanical Garden*; *Edgemere Landfill*, now a productive grassland and often overlooked; and *Flushing Meadows*, which

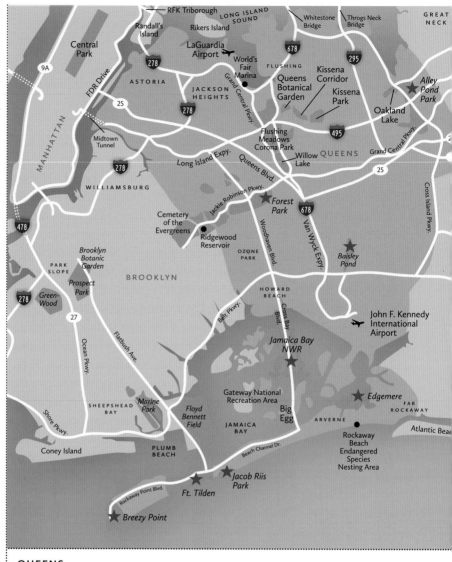

RFK Triborough
LONG ISLAND SOUND
Whitestone Bridge
Throgs Neck Bridge
GREAT NECK
Randall's Island
Rikers Island
Central Park
LaGuardia Airport
World's Fair Marina
678
295
9A
FDR Drive
ASTORIA
JACKSON HEIGHTS
Grand Central Pkwy.
FLUSHING
Queens Botanical Garden
Kissena Corridor
Kissena Park
Alley Pond Park
25
278
Midtown Tunnel
495
Flushing Meadows Corona Park
Oakland Lake
MANHATTAN
278
WILLIAMSBURG
Long Island Expy.
Queens Blvd.
Willow Lake
QUEENS
Grand Central Pkwy.
25
Jackie Robinson Pkwy.
Forest Park
678
Cross Island Pkwy.
478
Cemetery of the Evergreens
Ridgewood Reservoir
Woodhaven Blvd.
Van Wyck Expy.
Baisley Pond
Brooklyn Botanic Garden
OZONE PARK
PARK SLOPE
BROOKLYN
Prospect Park
HOWARD BEACH
278
Green-Wood
27
Belt Pkwy.
Cross Bay Blvd.
John F. Kennedy International Airport
Ocean Pkwy.
Flatbush Ave.
Jamaica Bay NWR
Edgemere
FAR ROCKAWAY
Shore Pkwy.
SHEEPSHEAD BAY
Marine Park
Floyd Bennett Field
Gateway National Recreation Area
Big Egg
JAMAICA BAY
ARVERNE
Atlantic Beach
Coney Island
PLUMB BEACH
Rockaway Beach Endangered Species Nesting Area
Beach Channel Dr.
Ft. Tilden
Jacob Riis Park
Rockaway Point Blvd.
Breezy Point

QUEENS

still has remnants of a bygone World's Fair. You'll also find short summaries of locations that are worth checking out in Other Places to Find Birds in Queens, of which *Baisley Pond Park* is our favorite.

If interesting and varied birding locations weren't enough, another great thing about birding in Queens is that while you can use a car to get around, pretty much everything is accessible by public transportation.

KEY SITES

Jamaica Bay Wildlife Refuge

Jamaica Bay Wildlife Refuge is one part of the vast nine-thousand-acre complex of preserves in the Gateway National Recreation Area. It's had an interesting ecological history, and prior to 2012's Hurricane Sandy was an unparalleled year-round birding preserve. Over the years, more than three hundred species of birds have been recorded at Jamaica Bay, and it remains the best of Queens birding. However, since the storm's breach of the West Pond, it's now mainly of interest for the quantity and quality of shorebirds in the fall and spring. The storm also brought saltwater to some of the upland areas, killing trees and affecting the songbird population. Until the replanting of those areas is completed and the breach is repaired, the refuge will host a reduced number of species. A scope is essential here, and mud boots, if you're planning to visit the East Pond. Poison ivy and ticks are always a hazard in this area during the summer, but at Jamaica Bay you'll have the privilege of visiting one of the few preserves in the New York City area where *quicksand* is a potential danger. But don't let this dampen your enthusiasm, as it is rare to be in a preserve that is so valuable for birds in such close proximity to New York City. A view of birds with the Empire State Building in the distance is an experience that just can't be beat.

VIEWING SPOTS

Cross Bay Boulevard divides the two main sections of the park. The visitor center is located on the western side of the street. The rangers on duty can supply you with a map of the park and a birding checklist, and will consult the tide tables if you haven't already. Out by the back door, take a look at the most recent notes in the wildlife log (housed in a brown wooden box).

Most people would now opt to cross the boulevard and explore the *East*

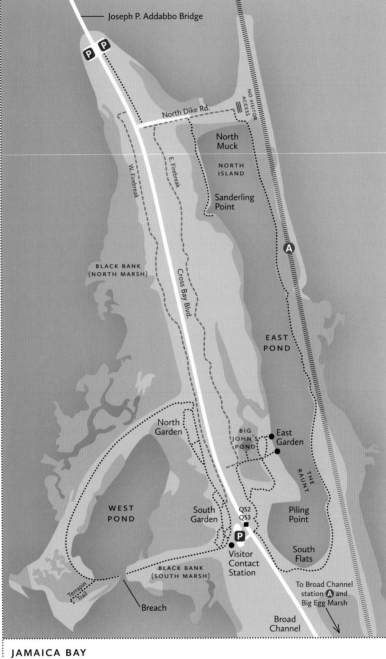

■ BUS STOP
⋮ TRAILS
| FIREBREAKS
||||| RAILROAD
🅿 PARKING AREA

Joseph P. Addabbo Bridge

🅿 🅿

North Dike Rd.

NO VISITOR ACCESS

North Muck

NORTH ISLAND

Sanderling Point

W. Firebreak

E. Firebreak

Cross Bay Blvd.

BLACK BANK (NORTH MARSH)

Ⓐ

EAST POND

North Garden

BIG JOHN'S POND

East Garden

WEST POND

South Garden

Q52
Q53

🅿

Visitor Contact Station

THE RAUNT

Piling Point

South Flats

Terrapin Trail

BLACK BANK (SOUTH MARSH)

Breach

To Broad Channel station Ⓐ and Big Egg Marsh

Broad Channel

JAMAICA BAY

American Avocets

Pond, saving a trip to the West Pond for later if they have the energy. If you're going to do the south end of East Pond, put on your mud boots and see the next paragraph. If you intend to skip that, leave your comfortable walking shoes on and take the trail to Big John's Pond (see below).

To bird the *south end of East Pond*, make an immediate right after crossing the street and move along the sidewalk, lugging your scope about a quarter mile to the trailhead. From here, pick your way through a jungle of phragmites to the flats. The National Park Service artificially lowers the water level to create mudflats in summer for shorebirds, so when the water level has been set properly, you can easily set up your scope on the exposed sand and get up close and personal with some very cool migrants like dowitchers and sandpipers. After spending some time here, unless you have plenty of time and stamina and have no fear of quicksand, resist the temptation to continue around the south side of East Pond along the mudflats flanking the train tracks. This is not a path we have ever taken, so our recommendation is to retrace your steps back to the sidewalk along Cross Bay Boulevard and take Big John's Pond Trail.

Big John's Pond is a little freshwater pond with a blind. It's all but hidden in phragmites, so keep a lookout for it on the left as you get close to the East

Pond. When you get near the turnoff, keep your group together and be really quiet as you approach and while in the blind. You'll be right on the edge of the water and will easily spook the birds with any noise. Across the pond and a little to the right you'll notice a Barn Owl nesting box. The owls there have been somewhat successful, but they are almost impossible to see unless you come at dawn and dusk. From Big John's Pond, make a left for a short walk to the western side of the East Pond. Essentially the same sort of birds that you will look for on the south side are here, but not as many shorebirds, and they're not as close. In season, you'll still see plenty of waterfowl, terns, and gulls.

Now reverse your path a very short distance and look for a trail on the right. It's a short walk to *another blind* on the East Pond and worth a visit. From there, retrace your journey back to the parking lot.

Return to your car, drive to the foot of the Cross Bay Bridge, and park in either parking lot. Leave your mud boots on, grab your scope, and walk down the eastern side of Cross Bay Boulevard on the sidewalk. After about one-third of a mile, you'll make a left onto an unpaved road. You can't park in here, but you can drop someone and the equipment off at the trailhead before parking in one of the lots at the foot of the bridge. From this unpaved road, you'll take the second trail on your right, the one through the trees, to the *northern side of the East Pond*. Here you'll find more mudflats with shorebirds eagerly feeding practically at your feet and in the distance. Once you're done here, it's not worth trudging up to the northeast corner of the pond, which is officially off-limits anyway.

If you want to visit the *West Pond*, return to your car and park again at the visitor center. Start your journey down the path behind the building. Check out the trees along the way, which may have birds of interest, but your main goal is to reach the tidal marshes and beaches. Here you'll look for shorebirds, waders, and waterfowl, plus terns and gulls. Jamaica Bay is home to several Osprey nests—one of which is very easy to see and photograph from the trail that goes to the breach at West Pond. (You can follow the Ospreys at http://www.jamaicabayosprey.org.)

Since, as of this writing, the breach along the southern portion of West Pond has not been repaired, this formerly brackish pond that provided for an additional number of species is now merely a saline inlet of Jamaica Bay. The 1.5-mile walk around the West Pond is no longer possible, nor are close views of birds on the far side, unless you trek around from the path on the right as you leave the visitor center. In the month of July particularly, and during high tide especially, be on the lookout at the breach for endangered Diamondback Terrapins, as they swim up to the beaches looking for a place to nest.

Walking back to the visitor center, you have the option of making a left and birding the *forested area between the West Pond and Cross Bay Boulevard* (the long path to the other side of the breach). It's best done first thing in the morning during spring and fall songbird migration. South of the visitor center, American Woodcock may be seen displaying in the *field* just before dark from mid-March through May.

KEY SPECIES BY SEASON

Spring Songbirds, including White-eyed and Blue-headed Vireo, Blue-gray Gnatcatcher; over two dozen species of warbler, including Common Yellowthroat, Black-and-white, Magnolia, Blackburnian, Black-throated Green, Wilson's, Blackpoll; shorebirds including American Oystercatcher, Black-bellied Plover, yellowlegs, Red Knot, Ruddy Turnstone, Dunlin; American Woodcock, Yellow and Black-billed Cuckoo, Ruby-throated Hummingbird, Northern Flicker; plus waterfowl including Ruddy Duck, Brant, Gadwall, Blue-winged Teal, scaup, Bufflehead, and many of the overwintering species.

Summer American Avocet, Snowy Egret; Great Blue, Little Blue, Green, and Tricolored Herons; Black-crowned and Yellow-crowned Night-Herons; Glossy Ibis, Osprey, Clapper Rail, American Oystercatcher, Black Skimmer, Willet, Hudsonian Godwit; Spotted, Solitary, Stilt, Pectoral, and White-rumped Sandpipers; Dowitchers; Least, Gull-billed, Common, and Forster's Terns; Willow and Great Crested Flycatchers; Tree and Barn Swallows; House, Marsh, and Carolina Wrens; Cedar Waxwing, Brown Thrasher, Common Yellowthroat, Yellow Warbler, Eastern Towhee, Baltimore Oriole, Red-winged Blackbird. Mostly, only common waterfowl remain in summer.

Fall Dunlin, Kildeer, Greater and Lesser Yellowlegs, Semipalmated Sandpiper, Belted Kingfisher; migrating songbirds including Eastern Phoebe, Northern Parula, Magnolia, Yellow-rumped, Palm, and Black-throated Blue Warblers (among many others); many sparrows, including Chipping, Savannah, Field, Song, Swamp, White-crowned; Purple Finch, Pine Siskin; Sharp-shinned, Cooper's, and Broad-winged Hawks, Merlin; some waders. Arriving waterfowl.

Winter Waterfowl include Snow Goose, America Wigeon, Northern Shoveler, Northern Pintail, Greater and Lesser Scaup, Canvasback, Common Goldeneye; Hooded, Common, and Red-breasted Mergansers, Pied-billed and Horned Grebes. Northern Harrier, Barn Owl (year-round), Golden- and Ruby-crowned Kinglets; American Tree, Fox, and White-throated Sparrows; Downy, Hairy, and Red-bellied Woodpeckers; American Woodcock.

HABITAT

Brackish and freshwater ponds, beach, mudflats, salt marsh, upland woods and fields.

BEST TIME TO GO

Spring through fall—fall shorebird migration begins in July, peaks in August and September, and continues through November. Songbird migration can be good here, as many species of warbler visit the area during peak months (April/May and September/October).

HOW TO GET TO JAMAICA BAY WILDLIFE REFUGE

By Subway A spur of the Ⓐ stops at Broad Channel. From there, walk to Cross Bay Boulevard and proceed north to reach the visitor center (about one-half mile) or East Pond trailheads.

By Bus The Q52 and Q53 make stops at the visitor center.

By Car This is the easiest way to reach Jamaica Bay and the surrounding parks. From the Belt Parkway, take the 17S exit to Cross Bay Boulevard South, and then stay on Cross Bay Boulevard to reach the north end parking lots or the visitor center. It is vitally important not to take the wrong Exit 17.

On your way home, you may accidentally wind up at Kennedy Airport while trying to find the Van Wyck. Don't beat yourself up about this; it's a rite of passage for every birder.

INFORMATION

Broad Channel, NY 11693
718-318-4340
General information: http://www.nyharborparks.org
www.nps.gov/gate/naturescience/upload/Jamaica-Bay-Bird-Checklist.docx
    can also provide a bird checklist
Andrew Baksh gives updates on his blog: http://birdingdude.blogspot.com
    Trails are open dawn to dusk. The visitor center parking lot is always open (but not the two lots at the north end near the foot of the Cross Bay Bridge). The visitor center itself is usually open from 8:30 a.m. to 5 p.m. It has the nicest restroom facilities and a wonderful little bookstore. You can

find portable toilets at the north end parking lots at the foot of the Cross Bay Bridge, and picnic tables at all lots. Food is not allowed on the trails. No pets.

Although it seems hard to believe, quicksand really is a problem here. Please note that to avoid sinking into deep mud, use only the West Pond trails, Big John's Pond Trail, and, at East Pond, step carefully on the mudflats and forgo a walk around the eastern side.

It could take several hours to properly explore Jamaica Bay, but if you aren't exhausted and in a car, you could reasonably combine this with a trip to one or more of the following: *Big Egg Marsh, Breezy Point, Jacob Riis Park, Fort Tilden, Edgemere Landfill, Canarsie Pier* and the *Brooklyn Coast, the Rockaways,* or *Floyd Bennett Field.*

You can expect ticks, so be prepared and dress accordingly—long sleeves, long pants, socks pulled over pant hems.

OTHER THAN BIRDING

Actually, apart from a nice walk through some wild places and a striking view of Manhattan, there is not much else going on. Bikes are not permitted, so if you bike out here, you need to leave it in the parking lot. During the full moon in May, Horseshoe Crabs come ashore to lay their eggs, and the rangers make an event out of it. Check out that and the other scheduled ranger-led activities at http://www.nyharborparks.org/visit/jaba.html. At nearby Marine Park you can rent a kayak and explore by water (http://www.wheelfunrentals.com). Entry into the salt marshes or onto the islands is not permitted.

Scope Highly recomended.

Photos Easy, but bring a long lens.

Big Egg Marsh, aka Broad Channel American Park

If you haven't spent the entire day at Jamaica Bay, this small inhabited island nearby has nesting birds and tidal marshes and great views of the Cross Bay Bridge and the Rockaways from the beach. It won't compare with Jamaica Bay, but if you want to be really thorough, and get a better view of the salt-water surrounding the preserve, it's minutes away and worth a quick stop.

VIEWING SPOTS

Park in the lot next to the paved athletic courts. Enter the field area and walk to the right along the fence (do not take the small path directly out of the parking lot, as it is loaded with poison ivy). Walk a short distance and find an opening in the old fence structure on the right. Follow this path out onto the marsh. This is a pretty view marred only by debris. Expect to see lots of Willets, Laughing Gulls, and nesting Osprey in season, along with Yellow-crowned Night-Herons. Listen for Clapper Rails. Boat-tailed Grackles and Tree and Barn Swallows are easily seen from the parking lot.

Returning through the fence to the ball fields, make your way out to the beach and the channel beneath the Cross Bay Bridge. There is trash here—from abandoned boats to fireworks. In winter there might be some waterfowl. It's recommended in summer and winter and only as part of a Jamaica Bay outing. It's not worth a special trip.

HOW TO GET TO BIG EGG MARSH

From Jamaica Bay Refuge, take the Cross Bay Boulevard south toward Rockaway Beach. The turnoff for the parking lot is less than a mile from Jamaica Bay and is on your right just before the bridge.

INFORMATION

Cross Bay Boulevard between West 20th Road and Beach Channel in Broad Channel

http://www.nycgovparks.org

During migration and in winter, the shorebirds and overwintering birds of Jamaica Bay are also seen here, but not as easily or in as great numbers. Take your scope.

Jacob Riis Park and Fort Tilden

For excellent fall and winter birding, visit these two contiguous Rockaway hot spots. Jacob Riis Park, envisioned as "the People's Park," is a popular summer beach and golfing destination. Fort Tilden is a 317-acre former army base that served to protect New York Harbor during the world wars. It is now

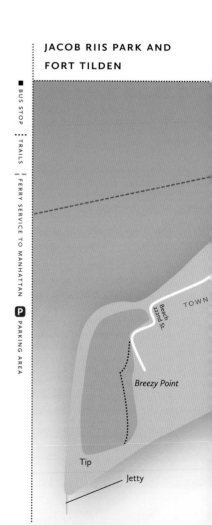

considered to have the premier hawk watch in Queens. Combined, these Rockaway spots offer a fairly interesting list of raptors, waterfowl, and, if you also visit in spring and summer, songbirds and shorebirds.

VIEWING SPOTS

At *Jacob Riis Park,* start by checking out the edges around the enormous *parking lot and the mall* behind the visitor center. In fall, your target will be sparrows. In winter, Horned Lark have been seen in flocks in the grassy areas, and look for Snow Buntings as well. Also try the *golf course edges* and take a swing past the *playgrounds and beach edges.* The winter waterfowl is fantastic. In summer, you can find beach-nesting birds like American Oystercatcher, but you'll have to contend with crowds of sunbathers.

To visit *Fort Tilden,* return to your car and drive back to Rockaway Point Boulevard heading west. Make a left turn onto Heinzelman Road and park in the large lot south of the ball fields. If the object of your visit is the *hawk watch* and nothing else, walk the road along the south edge of the ball fields,

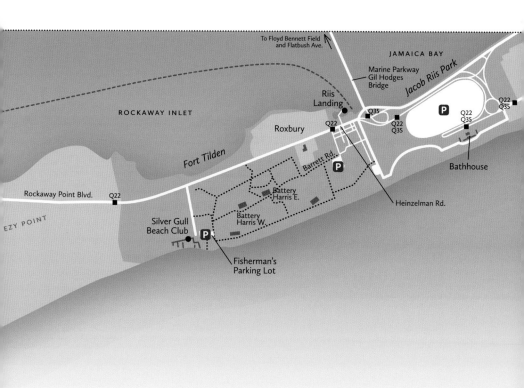

Laughing Gulls

Snow Buntings

Barret Road, toward the west as it gradually begins to taper off. After about a quarter of a mile, you'll reach the first of two World War II bunkers, *Battery Harris East*, which looks like a huge cement spaceship. Climb up to the top and enjoy the 360-degree views of the city, the Rockaways, and the ocean.

If you'd like to make a loop, upon leaving the Battery head south through the dunes to the beach, passing the *little pond*, which is often a good spot for migrants. Turn left at the water to get back to the parking lot. This little diversion is another opportunity to see migrating warblers, the occasional wader, and winter waterfowl.

Otherwise, on the way back to the main parking lot, the *vegetable gardens* and *grassy areas* along the north side may net some interesting grassland birds, including sparrows and maybe Horned Lark.

There is another hawk-watching location near the *bunker west of Battery Harris East*. If you have a fisherman's parking permit, you can park in the fisherman's parking lot at the far west end of Fort Tilden, but you still have a bit of a walk. You may want to check out the *beach*. In winter, interesting gulls, such as Bonaparte's, and several species of scoter may be seen. It would be helpful to have a scope.

From here, you can take the long hike farther west to another bird-attracting location, *Breezy Point*—see our description in this chapter.

KEY SPECIES BY SEASON

Spring Over twenty species of warbler, including American Redstart and Northern Parula; some shorebirds.

Summer Barn and Tree Swallows, Red-winged Blackbird, American Oystercatcher, Common Tern.

Fall Warblers returning south; Eastern Phoebe; American Pipit; Sparrows including Eastern Towhee, Field, Fox, Savannah, Swamp, White-throated, White-crowned. The occasional Vesper or Clay-colored Sparrow and other rarities have been seen. Kestrel, Merlin, Northern Harrier, Peregrine Falcon; woodpeckers including Northern Flicker.

WINTER Snow Bunting, Horned Lark; gulls including Bonaparte's; Purple Sandpiper; waterfowl including Horned Grebe, Long-tailed Duck, Common and Red-throated Loons; Northern Gannet, scoters.

HABITAT

Beach, edge, grassy areas, woods.

BEST TIME TO GO

Good year-round, but the most productive time is fall through winter when these two places are really active.

HOW TO GET TO JACOB RIIS PARK AND FORT TILDEN

By Bus Take the Q22 and Q35 to the Marine Parkway / Rockaway Point Boulevard stop.

By Ferry On weekends from Memorial Day to Labor Day you can take the Rockaway Beach Ferry from Pier 11 at Wall Street and the East River (http://www.newyorkbeachferry.com).

By Car Take Flatbush Avenue and Marine Parkway Gil Hodges Memorial Bridge to Rockaway Point Boulevard. Follow the signs to the Jacob Riis parking lot or continue on to Fort Tilden.

PARKING PERMITS AND RESTRICTIONS

Parking can be tricky here in the summer. It's a good idea to consult the Gateway Jamaica Bay Unit of the National Park Service (718-338-3799) to see what parking areas are open and what permits may be required at the time of your visit. At Jacob Riis Park, the main parking lot is open year-round; currently, there is a fee from Memorial Day through Labor Day from 9 a.m. to 6 p.m. If you come before or after collections each day, you can park for free. The rest of the year, the lot is free. From this lot, you can walk to Fort Tilden, although if the Fort Tilden parking lot is open it's more efficient to drive.

At Fort Tilden, the lots are subject to closure. If you have a fisherman's permit, you can park at the Fort Tilden fisherman's parking lot.

INFORMATION

Jacob Riis Park: 157 Rockaway Beach Boulevard, Rockaway Park, NY 11694
Fort Tilden: 169 State Road, Breezy Point, NY 11697
Jacob Riis Park / Fort Tilden Headquarters, 718-318-4300, www.nps.gov
/gate
Gateway NRA (NPS), 718-338-3799
http://www.nyharborparks.org

Restrooms are at the Bathhouse, Silver Gull Beach Club in the east, and between the two parks. There's no café in the off-season, but there are picnic tables near the beach. At Fort Tilden, the restrooms are near the ball fields.

Black Skimmers

OTHER THAN BIRDING

In the off-season, there may be some beachgoers, people fishing, hiking, and biking, plus others checking out Fort Tilden, which is listed on the National Register of Historic Places. Fort Tilden's Rockaway Artists Alliance offers art classes for adults and kids.

Scope Highly recommended.

Photos A long lens is very helpful.

Breezy Point

For nesting shorebirds in summer, migrating shorebirds in spring and fall, and views of a wide variety of waterfowl and some seabirds in winter, the westernmost point of land on the Rockaway Peninsula really delivers. In fact, Breezy Point hosts one of the two largest Black Skimmer colonies in New York—the other is seventeen miles away at Nickerson Beach, which we describe in the Nassau County chapter. This area is one of the wildest and least-visited of the beach preserves, and while it has excellent birding, it can be difficult to reach.

If you are looking for beach-nesting birds in late spring and early summer (April through June), head for the *the Tip*, where a highly managed site with roped-off areas protects nesting birds like endangered Piping Plovers, which can be reliably found along the beaches.

From July through August, the beaches just west of the Silver Gull Beach Club are colonized by nesting Least Terns and Black Skimmers. A battle is waged here every year between safe nesting for the birds and the encroachments of dogs, feral cats, and humans. It is essential to heed the stanchioned-off areas. To see American Oystercatchers and other beach nesters and their chicks, bring your scope. Check the shoreline for shorebirds; Roseate Terns have been seen here in spring.

In late fall through early spring (November through March), the westernmost tip of Breezy Point is also excellent for waterfowl and shorebirds. Near the *stone jetty* is a good place to set up your scope. Check the water for interesting ducks and seabirds like Northern Gannets and the occasional Razorbill. The jetty rocks can be a good place to find shorebirds feeding. From the *beach* you can also see migrating raptors in fall and through spring.

KEY SPECIES BY SEASON

Spring Over a dozen species of warbler pass through; shorebirds include Purple Sandpiper; huge numbers of Northern Gannets can be seen in April.

Summer Beach-nesting birds including Piping Plover, American Oystercatcher, Least and Common Terns, Black Skimmer. Spotted, Semipalmated, and Least Sandpipers make appearances.

Fall Migrating raptors including Northern Harrier, Sharp-shinned, and Cooper's Hawks; Ruddy Turnstone.

Winter Waterfowl including Long-tailed Ducks, Red-breasted Merganser, Red-throated and Common Loons; scoters, gannets, Horned Lark, and Snow Bunting. Razorbills have been recorded.

HABITAT

Beach, scrub, saltwater.

BEST TIME TO GO

This is a year-round destination, with special emphasis on nesting shorebirds in late spring and early summer; migrating shorebirds in spring and fall; and winter waterfowl.

HOW TO GET TO BREEZY POINT

The best way is to hoof it from Fort Tilden (see previous entry for directions), but it's a bit of a hike.

By Car Take Flatbush Avenue and Marine Parkway Gil Hodges Memorial Bridge to Rockaway Point Boulevard to its end at Beach 222nd Street.

Parking can be extremely complicated. On-street parking is nonexistent, the parking lots are small, and most require permits during the high season.

You can generally park at the Beach 222nd Street fisherman's parking lot in off-season months without a permit. Note that the distance to the Tip from here is about a mile, one way.

If you visit in-season you will need a fishing parking permit. Once parked, you can then walk west along the four-wheel-drive path. This parking lot is open year-round, but during the summer you need a special access pass for the day. The day pass can be obtained from the National Park Service (NPS) at Floyd Bennett Field. You must apply in person on the day you want the pass. NPS personnel are very friendly to birders and will try to accommodate you, but the desk is open only from 9 a.m. to 4 p.m. Even with a day pass, remember that the parking lot is relatively small. Calling ahead is advisable, as the requirements change from time to time.

Apply at Jamaica Bay Ryan Visitor Center at Floyd Bennett Field, Brooklyn, 718-338-3799, 9–4 daily, or e-mail GATE_JABAspecialparkuses@nps.gov.

INFORMATION

http://www.nyharborparks.org

You may be able to walk from the fisherman's parking lot to Breezy Point on the beach, or you may be stopped by security. Best to stick to the four-wheel-drive path, as noted above.

Fishing, beach walks.

    Scope Bring it.

    Photos A long lens is helpful.

Edgemere Landfill

Officially called the Rockaway Community Park, this former Superfund site is one of the oldest landfills in New York City. It dates from 1938, was closed in 1991, and it is now a gas reclamation project and infrequently birded area. Two gravel roads provide courses on different levels of this surprisingly attractive 173 acres of transitional grassland, waving meadows, and shoreline. Be prepared for grassland birds, songbirds, raptors, shorebirds, and stunning 360-degree views of Jamaica Bay, the New York City skyline, and JFK airport. Its proximity to the Jamaica Bay Wildlife Refuge makes it a recommended addition to a day of birding there or in the Rockaways at Jacob Riis Park, Fort Tilden, or Breezy Point.

VIEWING SPOTS

You might make a first stop at *the end of Beach 45th Street* to look out into the channel. Then drive into the main entrance at Beach 51st Street and Alameda Avenue and keep the Sanitation Department's buildings and salt depot to your right. To your left there will be *grassland*, to your right, edges and water. Keep the windows down and listen for birds, stopping when you hear or see activity. Stop frequently to listen and look for birds, especially on the right (water) side of the car. This road is not well traveled, so there should be no problem stopping on the side. When the road splits, stay right and take the lower road around the *base of the landfill*. Follow the lower loop as it curves to the left, and when JFK comes into view, stop the car along this straight stretch and look for shorebirds, sadly often found picking among the trash along the water's edge.

    After completing the lower loop, take the dirt road on the left to drive up and over the top of the hill, where you will be surrounded by a waving *grassland and meadow*. Here you can take in the amazing views of the water, the city, and the surrounding area. While there are no benches, you can park the car and have a nice tailgate break while watching Barn Swallows hunt dragonflies in summer over the meadows. Following that road will drop you off back on the lower drive. As you complete the circle again, make a right to enter the other part of the park, where you'll drive through a wooded area and come shortly to a parking area that gives access to two *wooden piers*. Join the fishermen on the piers and look around for ducks, cormorants, gulls, and wading birds like Black and Yellow-crowned Night-Herons, Snowy Egrets, and Glossy Ibis. In spring, be sure to check the wooded area for migrating songbirds.

In summer, you might want to check out the nearby *Rockaway Beach Endangered Species Nesting Area* (aka Arverne Piping Plover Nesting Area). Urban Park Rangers cordon off approximately eighty acres of shore between Beach West 38th Street and Beach West 58th Street for nesting Piping Plovers in this Forever Wild area. Least Terns and American Oystercatchers also take advantage of the protected habitat. To access the boardwalk, try parking on Beach 59th Street, south of Rockaway Boulevard near the playground. Once on the boardwalk, make a left to travel east. When you are watching the birds, take special care not to step on the highly camouflaged chicks or inadvertently separate the adults from the chicks, as doing so makes them vulnerable to gulls or other predators.

KEY SPECIES BY SEASON

Spring Edgemere is not as great a spot for migrant songbirds in spring as it is in the fall, but you will continue to see some waterfowl. Eastern Meadowlark migrate through; American Oystercatcher arrive.

Summer Nesting songbirds and grassland birds, Red-winged Blackbirds dominate; Bobolink, Common Yellowthroat and Yellow Warblers, Willow Flycatcher, Boat-tailed Grackle, Barn and Tree Swallows, Least and Common Terns, Willet, Savannah Sparrow, Osprey.

Fall Songbirds, including some warblers and sparrows, Eastern Meadowlark. Some raptors: Northern Harrier and Red-tailed Hawk are resident, but you could also see Cooper's and Sharp-shinned Hawks, Merlin and Peregrine Falcon.

Winter Snow Goose, Brant, Greater Scaup, American Wigeon, Bufflehead, Red-breasted Merganser, Common Loon, Great Cormorant.

HABITAT

Grassland, beach, edges.

Summer, fall, winter.

If you are making this part of a Jamaica Bay trip, you will need a car. The entrance is on Beach 51st Street and Alameda Avenue in Queens.

By Subway An Ⓐ spur stops at Beach 44th Street.

By Bus The Q22 and Q17 stop along Beach Channel Drive.

If you take public transportation, once there, be prepared for four to five miles of walking if you want to cover the entire area. Take along a bike for a nice biking-walking combo, and don't do it alone.

http://www.nycgovparks.org

No restrooms, no restaurant, no benches. Although it is now the Rockaway Community Park, this reclaimed landfill portion is not heavily visited except by fishermen and the occasional runner or biker. Not recommended if you are alone, even if you are traveling by car.

Despite the great views and attractive grasslands, there is trash along the shoreline, a not-so-gentle reminder that this site was once a dump.

Be aware that occasionally access is restricted.

Scope Would be a plus.

Photos Easy.

Forest Park

This woodsy park of over five hundred hilly acres is the largest forested area in Queens and is also a great place to see migrating birds. Located on the edge of the glaciation that created Long Island, its "knob and kettle" terrain accounts for the high ridges and gully depressions. Once home to the Lenape, Delaware, and Rockaway Native Americans, in the nineteenth century the land was assembled from a number of owners to become what is now Forest Park. Frederick Law Olmsted was involved in the park's plan and designed Forest Park Drive.

In spring and fall, its fairly dense eastern forest can be a migrant trap. Nearly every species of warbler and vireo visiting New York City has been seen

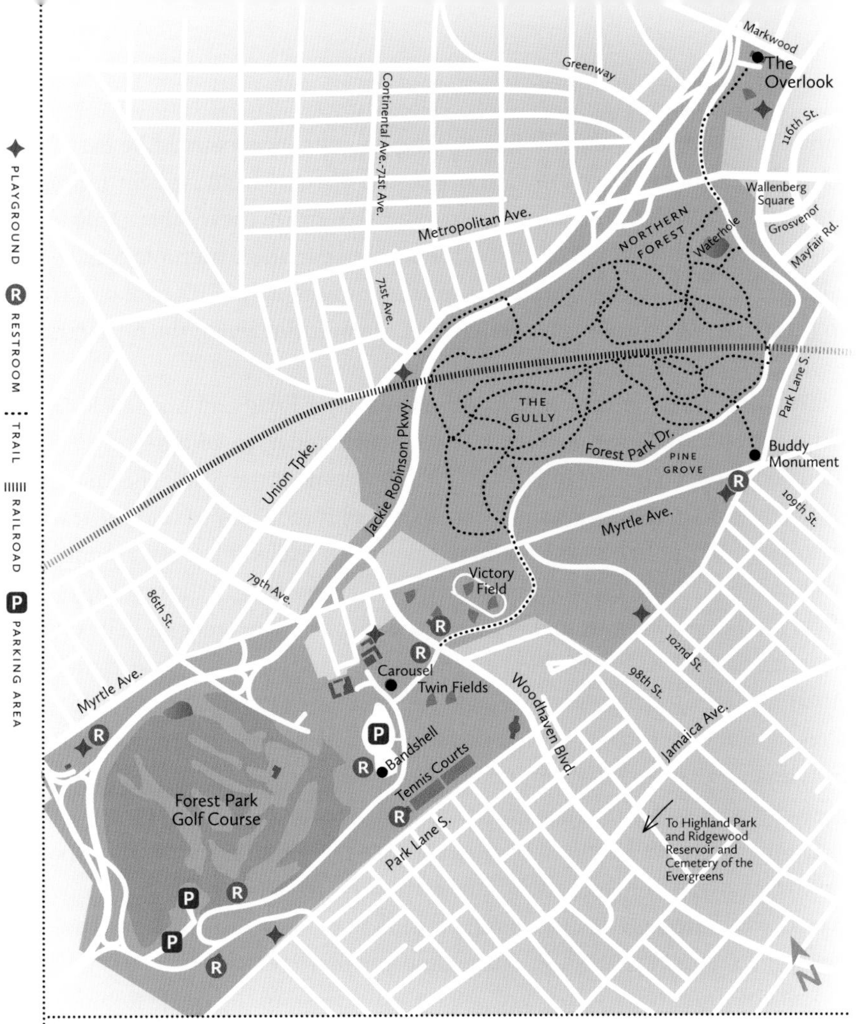

FOREST PARK

Map labels (clockwise/by area):

PLAYGROUND · RESTROOM · TRAIL · RAILROAD · PARKING AREA

Markwood · The Overlook · 116th St. · Greenway · Wallenberg Square · Grosvenor · Mayfair Rd. · NORTHERN FOREST · Waterhole · Continental Ave.–71st Ave. · Metropolitan Ave. · Park Lane S. · 71st Ave. · THE GULLY · Forest Park Dr. · PINE GROVE · Buddy Monument · 109th St. · Union Tpke. · Jackie Robinson Pkwy. · Myrtle Ave. · 79th Ave. · Victory Field · 102nd St. · 86th St. · Carousel · Twin Fields · Woodhaven Blvd. · 98th St. · Jamaica Ave. · Myrtle Ave. · Bandshell · Tennis Courts · Forest Park Golf Course · Park Lane S. · To Highland Park and Ridgewood Reservoir and Cemetery of the Evergreens · N

here, and it's no surprise. With a water hole and lots of mosquitoes, there is a lot to like if you're a songbird. And if you are looking for songbirds there is a lot to like, too—peaceful woodlands, spring migrants hanging from the branches over the Waterhole and lurking in the thick foliage. It's basically a treasure hunt, and you can easily get lost following fluttering birds and winding pathways in this wild and pretty spot.

VIEWING SPOTS

The park is bisected by Woodhaven Boulevard. The eastern part of the park is the wildest and has the most birds. West of Woodhaven Boulevard is where much of the human activity takes place, so your focus should be east of Woodhaven. If you have only a short period of time, head straight for *the Waterhole*, which is a little complicated to find but worth the trouble. Depending on the season and the rainfall earlier in the year, the Waterhole could be a large, mosquito-ridden pond or a small muddy slough. If it's cool out, try to time your Waterhole visit a bit later in the morning when the birds (and mosquitoes) will become more active.

To get there, enter the park between Grosvenor Road and Mayfair Road. Cross Forest Park Drive and look for two small boulders (medium-size rocks, actually) between lamppost N121 and N122 that mark the path. They might be partly obscured by vegetation, but it's the only sure way to find the trail. If you stay to the left on this path, in theory, it's really only a few minutes' walk to the Waterhole, if you know where you are going. But the first trip out there can be bewildering, so don't hesitate to ask a fellow birder for help. Our experience is that the birders in Forest Park are often locals and very friendly. Apart from the warbler-attracting habitat, there are often feeders as well. As a result, the Waterhole is the highlight of the park's birding; but if you have time and feel like roaming around, there are numerous trails throughout the park and a lake—all of which are scenic and pleasant.

If you plan to walk around the park, you might consider another hot spot called *the Gully*, which is across the railroad tracks, still in the eastern section, but west of the Waterhole. It's a glacial depression that holds freshwater and could be a good secondary stop if you have time and don't get lost trying to find it.

Still got time? Next stop is just a short drive west on the Jackie Robinson Parkway—*Highland Park* and the *Ridgewood Reservoir*—see our description in Other Places to Find Birds in Queens.

KEY SPECIES BY SEASON

Spring Nearly all species of Atlantic Flyway warblers are found here, including Cerulean, Bay-breasted, Hooded, Wilson's, Blue-winged, Worm-eating; White-eyed, Yellow-throated, Blue-headed, Warbling, Philadelphia, and Red-eyed Vireos; Great Crested Flycatcher, Blue-gray Gnatcatcher, Tree and Barn Swallows; Hermit, Bicknell's, Gray-cheeked, and Wood Thrushes; waders include Great Blue Heron, Great Egret, Green Heron; Belted Kingfisher.

Summer Eastern Wood-Pewee, Carolina Wren, Wood Thrush, Song Sparrow, Baltimore Oriole.

Fall Many warblers and vireos return in fall; American Goldfinch, Eastern Phoebe, Ruby and Golden-crowned Kinglet; Hawks include Cooper's, Red-tailed, and Broad-winged.

Winter Winter Wren; Sparrows such as Fox, Song, and White-throated, Dark-eyed Junco; Yellow-bellied Sapsucker and Downy, Hairy, and Red-bellied Woodpeckers.

HABITAT

Mature deciduous forest with many oaks; freshwater.

BEST TIME TO GO

Spring and fall for songbird migration.

HOW TO GET TO FOREST PARK

By Subway The area you want to visit is not particularly convenient to any subway station. Take the ⓙ to the 111th or 121st Street stations and walk north several blocks. You can also take the ⓔ and ⓕ to the Kew Gardens–Union Turnpike station and walk west on 80th Road.

By Train From the Long Island Rail Road's Kew Gardens station follow 83rd Avenue, turn right on Metropolitan Avenue, and continue for a few blocks.

By Bus Q11, Q23, Q37, Q53, Q54, and Q55.

By Car To the park's eastern side: from the Jackie Robinson Parkway take Exit 6, Metropolitan Avenue toward Kew Gardens. Park on or by Park Lane South. Or park near Myrtle Avenue in Richmond Hill.

INFORMATION

http://www.nycgovparks.org
http://www.nycaudubon.org

Once in the forest, it's easy to lose your way, so use your GPS or bring someone who knows the park.

Parking is along the streets. The parking lot is only for employees of the park itself. Don't find this out the hard way by being towed.

There are restrooms near the park office.

Scope Don't bother.

Photos Medium—there is a lot of foliage here in the spring and summer.

American Goldfinch on thistle

Queens Botanical Garden

A small, thirty-nine-acre gem of woodlands, wetlands, meadows, and lots of native plants and berry-producing trees, this pretty garden tucked into a Flushing neighborhood offers you an intimate experience. While not the birdiest of gardens, or as large or as well visited as the New York Botanical Garden in the Bronx or the Brooklyn Botanic Garden, it focuses on native plants in a variety of habitats attractive to birds. It's easy to navigate and on a scale that makes finding and viewing birds a cinch. You're unlikely to find crowds here, and after a short, relaxed walk, you can add on a trip to nearby Kissena Park or Flushing Meadows.

VIEWING SPOTS

Get a map at the entrance on Main Street, where you pay a small fee to visit, then make your own way around, paying particular attention to the areas with native plantings. These are especially attractive to birds, and you may find goldfinches right in front of you hanging on purple thistles in summer, and Chipping and Song Sparrows, Northern Mockingbirds, and Mourning

Doves close on the paths. Queens Botanical also has a bee garden and an abundance of flowers that hummingbirds love. The employees are helpful and friendly, and the garden offers occasional bird walks.

Try to make this trip spring through fall. In spring, there are Red-winged Blackbirds, songbirds, and sparrows migrating through; in summer expect American Goldfinch, Cedar Waxwing, Red-winged Blackbird; fall brings Red-tailed Hawk, Ring-necked Pheasant, Hermit Thrush; winter species are sparse, but look for Dark-eyed Junco and Mallard.

Getting photos is easy, and you don't need a scope. Most days there is a fee to enter.

HOW TO GET TO QUEENS BOTANICAL GARDEN

By Subway or Train Both the ❼ and the Long Island Rail Road (Port Washington line) make stops at Main Street/Flushing. From there pick up the Q44 or Q20 bus, or walk eight blocks south to QBG.

By Bus From the Bronx, Jamaica, or Flushing take the Q44 or Q20.

INFORMATION

43–50 Main Street, Flushing, NY 11355
718-886-3800
http://www.queensbotanical.org

There are restrooms and a gift shop in the main building. No café, but there are plenty of restaurants in the surrounding neighborhood.

Parking, both free and metered, can be found on the street, or for a day charge at a lot on Crommelin Street.

Alley Pond Park and Oakland Lake

Alley Pond, the second-largest public park in Queens, is a mishmash of 635 acres of reclaimed wetlands, beautiful forests with hiking paths, and recreational areas. The Alley Pond Environmental Center (APEC) manages the northern part of the park. It comprises the wildest and birdiest areas, including a salt marsh, which you can visit via the beautifully cared-for wooded hiking paths. In the 1930s this land was a dumping ground for debris from the construction of the Cross Island Parkway, and then, in the 1950s, for the Long Island Expressway. But don't let this deter you from a trip here. A

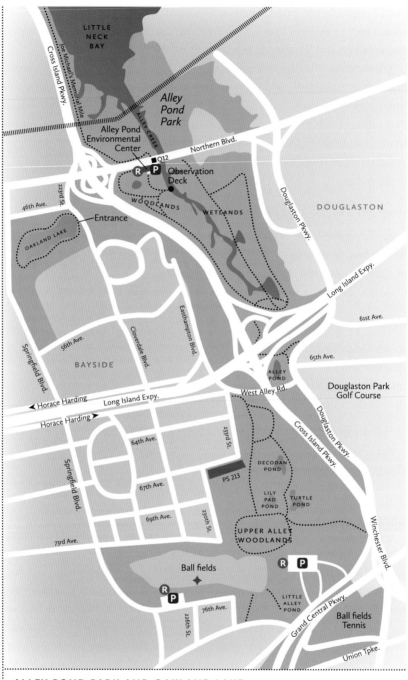

LITTLE
NECK
BAY

*Alley
Pond
Park*

Cross Island Pkwy.

Joe Michael's Memorial Mile

ALLEY CREEK

Alley Pond
Environmental
Center

■ Q12   Northern Blvd.

Ⓡ Ⓟ Observation
Deck

WOODLANDS

WETLANDS

DOUGLASTON

Douglaston Pkwy.

46th Ave.

223rd St.

Entrance

OAKLAND LAKE

Long Island Expy.

61st Ave.

56th Ave.

Cloverdale Blvd.

Easthampton Blvd.

BAYSIDE

65th Ave.

ALLEY
POND

Douglaston Park
Golf Course

Springfield Blvd.

Horace Harding ◄   Long Island Expy.

West Alley Rd.

Horace Harding ►

64th Ave.

233rd St.

PS 213

DECODAN
POND

Cross Island Pkwy.

Douglaston Pkwy.

67th Ave.

LILY
PAD
POND

TURTLE
POND

Springfield Blvd.

69th Ave.

230th St.

UPPER ALLEY
WOODLANDS

Winchester Blvd.

73rd Ave.

Ball fields

Ⓡ Ⓟ

Ⓡ
Ⓟ

✦

LITTLE
ALLEY
POND

226th St.

76th Ave.

Grand Central Pkwy.

Ball fields
Tennis

Union Tpke.

ALLEY POND PARK AND OAKLAND LAKE

wetlands reclamation project was started in 1972, and you can now see the amazing results. Walk the attractive bird-friendly habitat and search for the more than two hundred species that live in and visit the marsh and woodlands. Access to all the birding areas considered to be part of Alley Pond Park is made complicated by the transecting Cross Island Parkway from north to south, and the Long Island Expressway and Grand Central Parkway from east to west. All the pieces and parts of Alley Pond will be described here, but prioritize your visit to the area managed by APEC, as the real jewels are their wetlands and woodlands. If you have more time, check out the southern woodlands and Oakland Lake.

Great Blue Heron at Alley Pond Park

VIEWING SPOTS

The Alley Pond Park Environmental Center's *Alley Wetlands* and *Upper Alley Woodlands* have the best of the four birding locations. At the parking lot there is a visitor center that offers a variety of programs for families and children. Pick up a map there and go through the parking lot, taking the trail to the left. Follow it to the *boardwalk that borders the salt marsh*. Enjoy the great views from both the boardwalk and the pier that juts into the marsh. Native plants surround you, including cattails, and in summer you will see American Goldfinches at the thistles and Barn Swallows hunting low over the marsh. Red-winged Blackbirds breed here and fly among Snowy and Great Egrets, Killdeer, Black-crowned Night-Heron, and Belted Kingfishers. Nesting boxes placed in the marsh are popular with Tree Swallows. The *fields* bordering Alley Creek across from the pier offer nesting sites for Yellow Warbler, Baltimore Oriole, and Warbling Vireo. There can be a lot of avian activity here even as trucks whizz by just past the edge of the marsh on Northern Boulevard.

Returning from the pier, take a path on the left and enter the *wooded area*. During migration this area can be a migrant hangout, with three dozen species of warbler seen in these woods. The paths here are wide and well-maintained

QUEENS

111

and will take you past ponds where ducks, bullfrogs, and Black-crowned Night-Herons lurk. In summer, look for nesting Eastern Kingbirds, Gray Catbirds, and American Robins, which are pretty much everywhere. Eastern Wood-Pewees also nest here, as do Wood Thrush. Red-bellied, Downy, and Hairy Woodpeckers are year-round residents.

As you walk down the path, you eventually face a choice about the length of your hike and whether or not to continue around the pond. If you choose to circuit the pond, it can be a long trek without a lot of birds—especially if migration is over.

The above areas are the most scenic and most productive. If you want to visit other locations in the Alley Pond Park complex, the most interesting ones are *Oakland Lake* and the *southern woodlands* of Alley Pond Park below West Alley Road, described below.

However, from APEC, you could venture across Northern Boulevard to the Joe Riedl Wildflower Meadow salt marsh along the Joe Michaels Memorial Mile. But this is not the best birding area, and to continue north along the Cross Island Parkway is hardly an inspiring nature experience.

It might make more sense to drive to Oakland Lake across the Cross Island Parkway. The entrance to this lovely lake is on Cloverdale Boulevard, where you can park on the street. A path of approximately one mile traces the perimeter of the water. In spring, warblers can be found here. In summer, Barn Swallows flit over the surface of the water looking for bugs, and a pair of Mute Swans completes the lush picture. There may be nesting Warbling Vireos and Cedar Waxwings. In fall, before the water freezes, check for ducks, as you may find a diverse group that includes Lesser Scaup, Ruddy Ducks, Northern Pintail, and Pied-billed Grebe. In late winter, around March, the west end of the lake might have Rusty Blackbirds. The walk around the lake is pleasant, but be prepared for some wet and slippery terrain.

All in all, it's a nice walk, and it can be productive. It's definitely worth a try if you are in the area.

If you parked on Cloverdale, make a left onto 46th Avenue and then another left onto Springfield Boulevard, go under the Long Island Expressway, and make an immediate left onto Horace Harding. Follow Horace Harding to West Alley Road where it meets 233rd Street, and make a right onto 233rd. You are now at the side entrances to the woodlands in the *southern portion of Alley Pond Park*. Find on-street parking (there is a convenient entrance at the corner near the PS 213 playground) and start on one of the trails. If you're

Barn Swallow chicks at Alley Pond

not coming from Oakland Lake, use the parking lot on 76th Avenue between Springfield Boulevard and 226th Street. Southern Alley Pond's forest is full of birds in summer—from Scarlet Tanagers to Northern Flickers. The trails are well marked and meander past several kettle ponds. Mushrooms abound, as do mosquitoes and poison ivy. Following the main paths will take you in a loop back to where you entered.

The final stop is *Alley Pond* itself, uncomfortably wedged in among the Cross Island Parkway, the Long Island Expressway, and Douglaston Parkway. The entrance is on Douglaston Parkway, but park on one of the numbered side streets in the residential neighborhood. This body of water is also known as the Alley Park Restoration Pond, and if visited, it probably won't be the highlight of your day. In summer, the main delight could well be the Barn Swallows and their chicks nesting in the *tunnel*. In the winter, you may find interesting waterfowl, including Redhead, Bufflehead, and Wood Duck. As of this writing, it is not possible to walk around the lake near the water's edge, so good views of birds in the pond are often difficult to come by.

KEY SPECIES BY SEASON

Spring Three dozen species of warblers, including Blue-winged, Bay-breasted, Blackburnian, Blackpoll, Chestnut-sided, Black-throated Blue, and Black-throated Green; White-eyed, Yellow-throated, and Blue-headed Vireos; Tree Swallows also nest here; Blue-gray Gnatcatcher, Rusty Blackbird.

Summer Osprey; Great and Snowy Egrets, Black-crowned Night-Heron, Killdeer, yellowlegs, Spotted Sandpiper, Laughing Gull, Belted Kingfisher, Chimney Swift, Barn Swallow, Eastern Wood-Pewee, Willow and Great Crested Flycatchers, Eastern Kingbird, Wood Thrush; Warbling and Red-eyed Vireos; Cedar Waxwing, Scarlet Tanager, Baltimore Oriole, Yellow Warbler, American Goldfinch, Song Sparrow, Rose-breasted Grosbeak, Carolina Wren, Red-winged Blackbirds.

Fall Returning warblers, Golden and Ruby-crowned Kinglets, Hermit Thrush, Eastern Towhee, Swamp and Fox Sparrows, Pine Siskin, Sharp-shinned and Red-tailed Hawks, Great Horned Owl; Hairy, Downy, and Red-bellied Woodpeckers year-round.

Winter Oakland Lake has waterfowl that include Gadwall, Wood Duck, American Black Duck, Northern Shoveler, Northern Pintail, Redhead, scaup, Ring-necked Duck, Ruddy Duck, Pied-billed Grebe.

HABITAT

The northern part of the park is the most productive, and it's composed of well-managed tidal wetlands and woodlands in extreme close proximity to and transected by major highways. The southern portion is made up of forest and kettle ponds. Other locations in Alley Pond offer shoreline and freshwater ponds, as well as woodlands.

BEST TIME TO GO

The best times to go, in order, are spring, fall, summer, and winter.

HOW TO GET TO ALLEY POND PARK

APEC is located at 228–06 Northern Boulevard, Douglaston. It's just east of the Cross Island Parkway on Northern Boulevard in northeast Queens—a one-story green building with parking behind the center.

By Subway and Bus Flushing Line **7** to its last stop, Flushing–Main Street. Transfer at Roosevelt Avenue to the Q12 bus heading east. It will eventually run along Northern Boulevard. APEC has a bus stop right in front of the center.

By Train The Long Island Rail Road stops at Bayside (Port Washington line), where you can walk south on Bell Boulevard to Northern Boulevard and catch the Q12 bus.

By Car If your GPS doesn't give you the right directions using Douglaston as the town, try using Little Neck instead.

228–06 Northern Boulevard, Douglaston, NY 11362
718-229-4000
http://www.alleypond.com
http://www.nycgovparks.org

Restrooms are found in the Alley Pond Park Environmental Center. No restaurants.

Be aware that poison ivy is present generally on the sides of the trails; mosquitoes and ticks can also be present.

Scope Not needed.

Photos Easy.

OTHER PLACES TO FIND BIRDS IN QUEENS

There are a number of parks in Queens that can be interesting to visit and which play host to a good variety of birds. None of these individually rank as very birdy, but they have their followers, and if you are looking for a new place to discover, any could make for a nice outing. The following is a list of the better ones—the best sites are marked with an asterisk.

Baisley Pond Park*

Just north of John F. Kennedy Airport in South Jamaica, and next to the Baisley Housing Project, lies an interesting park that is somewhat off the radar. Particularly appealing to photographers in winter, its main feature is the freshwater pond, which sustains overwintering waterfowl. Even when most of it freezes, there are still areas of open water where Redhead, Ring-necked Ducks, Gadwall, Red-necked Grebe, American Wigeon, Northern Shovelers, Ruddy Ducks, and other waterfowl are pressed toward the shoreline, making for easy close-up photographs. Wood Ducks also visit but are less willing to pose. It may be the best spot in Queens to see Redhead, and there are gulls galore—including banded gulls. Enter this urban park by parking in the lot on Baisley Boulevard, across from August Martin High School. Take the loop walk around the park. If you still have time, the cricket fields south of Rockaway Boulevard host Horned Larks and geese in winter. Scan the grass and trees for sparrows and woodpeckers in fall. Highly recommended in winter, this can be a pleasant year-round stop, but definitely best done with a pal.

Rockaway Beach Endangered Species Nesting Area*

This site is also known as the Arverne Piping Plover Nesting Area, and the beach is dedicated to protecting Piping Plovers, Least Terns, and American Oystercatchers. Located between Beach 45th Street and Beach 56th Street, this is a late spring through summer spot to visit when the birds are raising their chicks. You are going to be glad that you brought your scope, as the viewing is from afar so as not to disturb the birds, which are protected by the Urban Park Rangers. To get there, take the Ⓐ to the Beach 44th station, then look for the signs. http://www.nycgovparks.org. For more access information see our description earlier in this chapter under Edgemere Landfill.

Kissena Park and Corridor

Not far from LaGuardia Airport lies Kissena Park, the main attraction in the several-mile-long Kissena Corridor Park and part of the Brooklyn-Queens Greenway. These parks were part of a nineteenth-century railroad right-of-way, and the raised nature trail was the former main line of the Central

Piping Plover chick

Railroad of Long Island. Now several parks have been combined into an open, highly maintained landscape with lakes, some mature trees, and a marsh. With such a varied habitat, the list of birds seen here over time is impressively large—almost two hundred species as of this writing—but on any given day, the birds are not as plentiful as you would find at Alley Pond or Forest Park. If you are in the neighborhood, however, take advantage of the waterfowl on the lake in winter, songbirds during migration, and some interesting nesting birds in the summer. Enter the park at Rose Avenue and Parsons Boulevard. The Korean War Monument is just ahead, and near there, a restroom. Otherwise, proceed north and left to Kissena Lake to explore it and the trees on the hill adjacent. Follow this with a walk through the marsh to the south.

The Kissena parks are north of the Long Island Expressway and east of Interstate 678. They are not easily reached by subway, but the Q65, Q17, Q25, and Q34 buses will put you on their doorstep. If you take the ❼ to its terminus at Flushing-Main Street, you can pick up any of these buses there.

Kissena Corridor has some open, brushy areas that are best visited in fall during sparrow migration, when Indigo Buntings come through as well.

The parks are nearly walking distance from the Flushing-Main Street stop on the ❼, where you can pick up the Q17, Q25, Q34 or Q44.

Flushing Meadows Corona Park

Flushing Meadows Corona Park emerged from the Corona ash dumps to become grounds for the 1939/1940 and 1964/1965 World's Fairs. Now it is a large park wedged between highways and retains such vestiges of those bygone days as the iconic Unisphere. Flushing Meadows also has an impressive life list, owing to its varied habitats of freshwater lake, open lawn, and stands of trees. Again, this is not an outstanding birding destination, but if you are in the area, it might be worth the stop. For winter waterfowl, make Meadow Lake a priority. Park near the boathouse and also check out the creek that feeds Meadow Lake from the north and runs beneath the highway. Continue north to the man-made pond at the Promenade of Industry and finish in the far north at the pond near the miniature golf course. You'll also get a few migrants spring and fall and a sparse number of nesting birds. The Red-tailed Hawks nesting in the Unisphere could be circling anywhere at any time. To see the nest, aim your scope on the metal equatorial band below India.

Flushing Meadows Corona Park is south of the Long Island Expressway and lies on the west side of Interstate 678. It is a heavily used public park with plenty of restrooms and interesting things for the non-birder. In addition to all the usual outdoor recreation, it hosts dragon-boat races and has a magnificent stand of blossoming trees near the Westinghouse Time Capsule. The ❼ to the Mets–Willets Point stop will deliver you to the northern part of the park, where you can take the above-described route in reverse.

Willow Lake

Willow Lake, just south of Meadow Lake, is seldom visited and is not accessible from Flushing Meadows Corona Park. You'll have to leave that park and enter from Park Drive East and 73rd Terrace. A pedestrian bridge takes you over the Van Wyck Expressway and eventually to the water. The best time to go, for interesting birds, is in the winter, but you don't want to do this alone.

World's Fair Marina*

World's Fair Marina, located on Flushing Bay across Northern Boulevard from Shea Stadium and Flushing Meadows Corona Park, can be worth a visit for winter waterfowl. Look for Canvasback, Redhead, and Red-breasted Merganser, along with more common Bufflehead, scaup, and the like. It's a little tricky to get to and is best reached by car: 125–00 Northern Boulevard, off the Grand Central Parkway, Flushing Bay; 718-478-0480.

Highland Park and Ridgewood Reservoir

Built in the nineteenth century and straddling the Queens and Brooklyn border, Ridgewood Reservoir has commanding views from the high ridge whence its name derives. It's a good spot to look for raptors overhead, and the reservoir basins can have some of the more common waterfowl. Occasionally, a Wood Duck finds it appealing. The wooded areas in Highland Park have resident woodpeckers and some interesting nesting birds. During migration there can be over two dozen warbler species. Try the Ridgewood Reservoir Trail between basins 1 and 2 for some of the more than 150 birds found here. There are athletic fields in the south, and the Brooklyn-Queens Greenway runs into the park and around the reservoir. The park is located between Cemetery of the Evergreens and Forest Park.

Blue Jay

Cemetery of the Evergreens

Cemetery of the Evergreens (aka the Evergreens Cemetery) is one of a few cemeteries to the west of Highland Park and across Jackie Robinson Parkway. Technically, it's in Brooklyn, but since it's nearly contiguous with Highland Park, we included it here. While it might not provide much for your life list, it's a peaceful spot, and birders are encouraged to visit. Enter off Bushwick Avenue. Reached by many subway lines, Ⓐ Ⓒ Ⓙ Ⓩ and Ⓛ at the Broadway Junction stop.

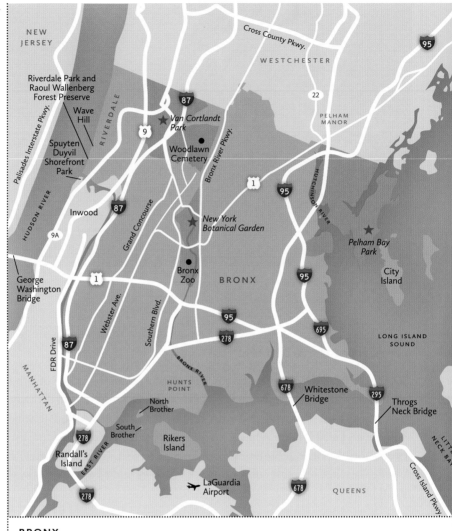

NEW
JERSEY

Cross County Pkwy.

WESTCHESTER

95

Riverdale Park and
Raoul Wallenberg
Forest Preserve

Wave
Hill

87

22

PELHAM
MANOR

RIVERDALE

★ Van Cortlandt
Park

9

Spuyten
Duyvil
Shorefront
Park

Bronx River Pkwy.

Woodlawn
Cemetery

HUTCHINSON RIVER

Palisades Interstate Pkwy.

Inwood

87

1

95

HUDSON RIVER

9A

Grand Concourse

★ New York
Botanical Garden

★ Pelham Bay
Park

City
Island

George
Washington
Bridge

1

Bronx
Zoo

BRONX

95

Webster Ave.

Southern Blvd.

95

695

LONG ISLAND
SOUND

FDR Drive

87

278

BRONX RIVER

MANHATTAN

HUNTS
POINT

678

Whitestone
Bridge

295

Throgs
Neck Bridge

North
Brother

LITTLE NECK BAY

278

South
Brother

Rikers
Island

EAST RIVER

Randall's
Island

✈ LaGuardia
Airport

678

QUEENS

Cross Island Pkwy.

278

BRONX

4 The Bronx

The city's northernmost borough was founded in 1639 and named after its first European settler, Jonas Bronck, who, uncharacteristically for New York City, was not Dutch, but Scandinavian. Although this area is heavily populated and thoroughly urban, it still retains nearly 24 percent of its forty-two square miles in parkland, interspersed with a dozen universities and colleges and bucolic residential areas such as Riverdale. There are some terrific places to bird here, but for safety reasons, in nearly all of them we recommend that you do not make the trip alone.

The best place to bird in the Bronx is *Pelham Bay Park*, the largest park in New York City. It has enough contiguous land and water and is wild enough to support more than two hundred species of birds. Pelham is the prime spot, but don't pass up other great opportunities like the *New York Botanical Garden*, *Woodlawn Cemetery*, and *Van Cortlandt Park*—another huge park of over one thousand acres. We also provide a list of smaller parks that may be less productive but are worth a visit if you're in the neighborhood.

KEY SITES

Pelham Bay Park

This is actually New York City's largest park, and at nearly three thousand acres it is over three times the size of Central Park. This park has been designated an Important Bird Area, and it has a variety of habitats, including some 800 acres of forest and 350 acres of salt marsh or flats. In summer, there is a lot of human activity, so keep this in mind when planning a visit. Formerly inhabited by the Siwanoy tribe, then the Dutch, it's had more recent use by the military, fire department, and police for training.

Now the park is known for the attractive historic Bartow-Pell Mansion,

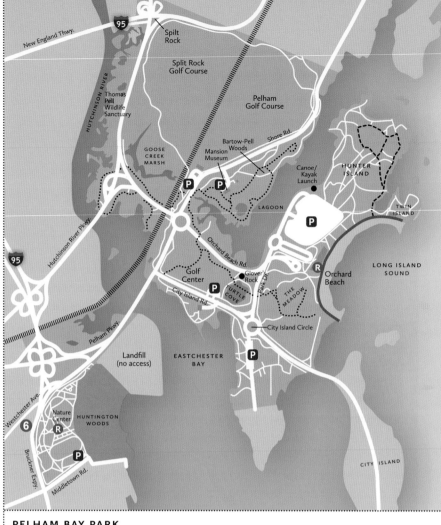

PELHAM BAY PARK

thirteen miles of man-made beaches (including the only public beach in the Bronx), two golf courses, hiking and biking trails, all kinds of ball courts, and lots of other things to do. It's a mixed-use park and is a popular spot for local residents to do pretty much everything other than birding. There are really only a few good spots to find birds, but those spots are superb. And while birds can be found year-round here, you may find the less-visited winter season a better experience.

VIEWING SPOTS

122   All signs in the park direct you to or away from *Orchard Beach*, "the Riviera of the Bronx." It's a huge semicircular man-made sand beach that attracts

thousands of visitors a year, which is the reason that we recommend Pelham Bay in its off-seasons. Nevertheless, it is the park's major landmark, has the biggest parking lot, and is good for orienting yourself to the better birding locations.

The most attractive off-season is winter. Pelham Bay is famous for owls, and Great Horned, Long-eared, and Northern Saw-whet Owls are all seen here. The best way to go about your owl quest is to take a guided walk with a park ranger or with someone who gives regular walks in the park. They can lead you to the favored roosting or nesting spots and give you tips about how to find them when you come alone. If you come on your own to Pelham Bay, the best place to look for owls is Hunter Island and the woods around the Bartow-Pell Mansion. And even if you don't find an owl, you will be rewarded with sightings of interesting waterfowl and, as winter gets fully under way, Harbor Seals in their haul-outs. The best place to view seals is at the northern end of Orchard Beach's boardwalk, but they can occur anywhere on the open water around Hunter and Twin Islands.

Otherwise, prioritize your visits to various locations depending on the season. Since you could be reading this at any time of the year, let's begin our tour with the area in the northeast section known as *Hunter Island*, though *Twin Island* is also good at the same times of year. Hunter Island is where to

Palm Warbler

go for woodland birds and, along its shores, waterfowl. Park in the parking lot for Orchard Beach in its northeast corner, and access the trail as shown on the map. On Hunter, it doesn't matter whether you stick to the *Kazimiroff Nature Trail* or go off onto the other trails covering the area—follow your instincts and you'll come across migrating songbirds, wading birds, and overwintering waterfowl. However, you'll miss a transporting experience if you do not hike along the eastern, water-facing edge of Hunter and Twin Islands for a mini-vacation to Maine's rocky coast.

Another area to try is the large, artificially enlarged body of water known as *the Lagoon*. The best and clearest access is afforded at the canoe launch in the northwestern corner of the parking lot. In winter, you'll see terrific waterfowl. In fall, the grassy area just before the break in the phragmites is a good place to stop and look for migrating sparrows. Low tide is the best time to come. You can get out on the pebbly beach, but be careful—it's slippery, and the exposure is going to last only a few hours. This is also a great spot to see migrating Osprey—especially in the fall. Bring your scope.

Across the Lagoon lies the historic *Bartow-Pell Mansion*. The mansion itself and the gardens are not elaborate, but history buffs won't want to miss them. Park in the lot and scan the lawn for migrating sparrows in season, nesting birds in summer. Check the shrubs around the mansion for roosting Northern Saw-whet Owls, who seem to like the area. Explore the *woods* near the mansion for songbirds and take *the trail along the Lagoon* for waterfowl. There are several openings in the phragmites to look out onto the water, and these openings will get more frequent as you travel north. Don't forget to look to your left at the marshland—something good may turn up. At a certain point on the trail, you'll come to a stile. Unless you want to take a long hike, do not take the Siwanoy Trail. Instead, make a sharp left back to the mansion, effecting a loop.

Another often productive spot is *Turtle Cove*. We hesitate to mention it because you're not actually supposed to park in the lot if you're not using the golf center, but during the off-season, there are plenty of spaces. If you're coming by bus, there's a convenient stop at City Island Circle. From the lot, you'll see a *wood-chip path* that runs along the driving range. That will take you along the border of some woods, which are worth checking out. As you get to the end of the driving-range fence, a trail will open on the right. Go through the woods, coming out on the northern side of Turtle Cove—a tidal, saltwater inlet that can have wonderful waterfowl and wading birds.

The open areas around here host migrating sparrows in the fall. Continue your loop around the pond, and if you like, cross City Island Road and look into the open water of Eastchester Bay. Full circle is about a one-mile hike.

While *the Meadow* is not usually especially birdy, it is one of the few places in New York City to observe American Woodcock in breeding display. Go in March or early April and use the rocky ridge on the east side at dusk. You'll need special permission to be in the park after dark. (See Information below for details.)

For birdwatchers, those are the best parts of Pelham Bay, with the following provisos: (1) At this writing, the southwestern portion of the park, known as Pelham South and the portion accessible by subway, is under renovation. Part of it was a wooded area called Huntington Woods and, formerly, Huntington Estates. When the renovation is completed it may once again be a productive woodland with access to Eastchester Bay. For now, it is closed to visitors. (2) The landfill northeast of Pelham South is off-limits. The only way to get there is to park in the lot at the southern end of the park off Middletown Road (or along the street near the playground), walk the mile and a quarter to the southwestern edge, and look in through a chain-link fence. This is unfortunate, as the area is a wonderful habitat for grassland birds, and the site itself will one day make for a terrific hawk watch. We hope that by the time you are reading this, the landfill will have come into its own as a nature preserve and be appropriately accessible to only passive recreation.

If you are looking for birds, we can't really recommend that you bother with the golf course, Goose Creek Marsh Trail, Rodman's Neck, or any other sections of the park, although we've always wanted to have enough time to hike out to Split Rock. The Siwanoy Indians considered it a sacred space, and it has become a colonial history landmark as well—it's where Anne Hutchinson, expelled from the Massachusetts Bay Colony, was killed by the Siwanoy in retaliation for the treatment they were receiving from the Dutch.

KEY SPECIES BY SEASON

Spring Migrating raptors and songbirds, including Eastern Towhee; Pine, Yellow, and Black-and-white Warblers; American Woodcock, Greater and Lesser Yellowlegs, Northern Flicker, Tree and Barn Swallows. Clapper Rail will persist through the breeding season.

Summer Osprey, wading birds, Willow Flycatcher, Warbling and Red-eyed Vireos, Marsh Wren, Yellow Warbler, American Tree Sparrow, Red-winged

Blackbird, Orchard and Baltimore Orioles, Yellow-crowned Night-Heron, Great and Snowy Egrets, Wild Turkey.

Fall Migrating Osprey; Chipping, Field, Savannah, Nelson's, Fox, Swamp, White-throated, and White-crowned Sparrows; Purple Finch, Pine Siskin, Golden-crowned and Ruby-crowned Kinglets, Hermit Thrush; two dozen species of migrating warblers; migrating raptors including Red-shouldered Hawk and Northern Harrier; Indigo Bunting, Cedar Waxwing.

Winter Brant, Gadwall, Canvasback, scaup, Bufflehead, Common Golden-eye, Hooded and Red-breasted Mergansers, Horned Grebe, Great Cormorant, Red-throated and Common Loon, American Wigeon, with the occasional Eurasian Wigeon. Great Horned Owls are year-round, but easier to see this time of year. Also possible are Barred, Long-eared, and Northern Saw-whet Owls.

HABITAT

Grassland, old-growth forest, salt marsh, open saltwater, and thirteen miles of shoreline.

BEST TIME TO GO

Best and least-busy time to go is winter, but birds can be found year-round here.

HOW TO GET TO PELHAM BAY PARK

By Subway Take the ⑥ to the last stop, Pelham Bay Park. From there, you'll be in the southern portion of the park and can take the Bx29 as far as the Turtle Cove area and get off at the City Island Road / Shore Road stop. From here it's about a one-mile walk to the Bartow-Pell Mansion or to Orchard Beach.

By Bus The Bx5 and Bx12 serve Orchard Beach but only during the summer.

By Car Set your GPS to Orchard Beach, Pelham Bay Park. You should be instructed to use I-95 and take exit 8B to Orchard Beach / City Island, then follow the signs to the Orchard Beach parking lot. If you would prefer to go straight to Turtle Cove, turn right at the first traffic light after crossing the Hutchinson River drawbridge, and look for the sign for the Turtle Cove Golf Center on the left.

INFORMATION

Administrative Office: 718-430-1891

http://www.nycgovparks.org

Friends of Pelham Bay Park: http://www.pelhambaypark.org
Debbie Becker often writes about Pelham Bay on her website,
    http://www.birdingaroundnyc.com. So does Jack Rothman,
    http://cityislandbirds.com
http://www.nycaudubon.org

Restrooms, noted on the official map as all over the park, are mostly closed in the off-season, but those in the Orchard Beach Pavilion are open year-round.

OTHER THAN BIRDING

You can see Harbor Seals in winter. There is fishing, golf, baseball, basketball, track, biking, bocce, football, handball, horseback riding, roller hockey, soccer, kayaking/canoeing, swimming, excellent playgrounds, tennis, hiking (try the 3.5-mile-long Siwanoy Trail). Lovers of history can visit the Bartow-Pell Mansion Museum and notable spots such as Split Rock and Glover's Rock. For geologists, Pelham Bay Park represents the southern terminus of New England's rocky coast. Glacial action has exposed feldspar, garnet, schist, and quartz and has left erratic boulders.

The park has a full range of special events that include yoga, art lessons, Horseshoe Crab outings, and survival skills training.

If you've come by car, you may want to have lunch on City Island. It's a fascinating and unique piece of New York City that has plenty of restaurants overlooking the water. For picnicking, make your way to the rocky coast of Hunter or Twin Islands or, if more convenient, to the large flat rock by the canoe launch with views of the Lagoon.

You must pay for parking in the summer, but it's otherwise free.

Scope Would be helpful.

Photos Medium difficulty.

New York Botanical Garden and Bronx Zoo

These two New York City institutions, both founded in the nineteenth century, share adjoining parkland in the center of the Bronx bisected by the Bronx River but provide completely different experiences.

The Botanical Garden, a Historic Landmark, was founded in 1891 and is an inviting urban oasis. With 250 acres of varied natural terrain, including dramatic 250-year-old forests, with hills, waterfalls, and ponds, this can be a

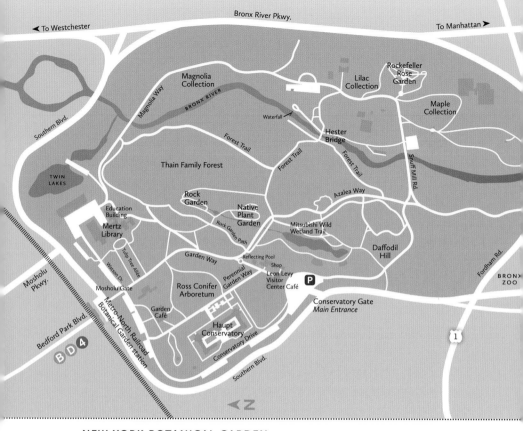

NEW YORK BOTANICAL GARDEN

good birding destination during migration. It also is a nesting area for Wood Ducks and offers the possibility of seeing their ducklings in May. If you enjoy horticulture, you're in for a double treat.

The Bronx Zoo, a part of the Wildlife Conservation Society, is, at 265 acres, one of the largest urban zoos in the world. It opened in 1899 and now houses over four thousand animals representing 650 species—many in fairly natural settings.

While the Bronx Zoo is probably the better known, you are going to have a more productive birding experience at the Botanical Garden. And because both of these places are well mapped, well traveled, and secure, they make great places to go if you want to bird alone. The best birding sites are in the New York Botanical Garden or along the Bronx Zoo's Jungle World Road and Mitsubishi Trail.

Enter the New York Botanical Garden through the main entrance, the Conservatory Gate, off Southern Boulevard. Pass through the ticketing center and make your way to the Reflecting Pool. From there, turn right and follow the trail to the beautiful *Thain Family Forest*, the largest uncut old-growth forest in New York, famous for its year-round Great Horned Owls. You can do a loop by taking either of the two trails that come up on your left. The first is a shorter route, but you'll go through more interesting terrain by taking the second left. This trail eventually brings you up along the *Bronx River*, often a good spot, and to *Magnolia Way Road*. Make a right there in order to look out over the river again from the bridge. Double back on Magnolia and explore the *Twin Lakes* area, which is always productive. Take any trail back toward the Reflecting Pool, but stop in the *Native Plant Garden*, especially in fall. Lastly, take in the short but beautiful *Mitsubishi Wild Wetland* Trail, which gives you the chance to check out the ponds and trees for activity.

If you are up for a birding/animal-viewing adventure, take your binos to *the Bronx Zoo*, which borders the south end of the Botanical Garden. Despite its exotic inhabitants, it has a decent list of native birds—especially waterfowl, woodpeckers, swallows, fall and winter sparrows, and year-round common residents. However, there are warblers to be found here as well during migration, and you can also find interesting summer nesters like Baltimore Oriole.

Park at the Bronx River Gate, where you will pay a fee, and bird along *Jungle World Road*. From the parking lot, cross the footbridge toward the ticket booth, and before you reach the booth, make a left on the wooded road that follows the river. It's a nice place for migrating songbirds.

Or, as you're leaving the parking lot, you could bird the *Mitsubishi Riverwalk*, a restored ecosystem. Instead of crossing the footbridge, find the restrooms at the parking lot and make a right, then curve around to the left to start the trail. This route is also good for migrating songbirds, and since the Mitsubishi Riverwalk was engineered to provide nice views of the river and its waterfalls, this will be the more scenic and potentially more productive choice.

HOW TO GET TO THE BOTANICAL GARDEN

By Subway Take the Ⓑ Ⓓ or ④ to Bedford Park Boulevard station and either transfer to the Bx26 bus east for the Mosholu Gate entrance or walk eight blocks on Bedford Park Boulevard. Turn left onto Southern Boulevard and walk one block to the Mosholu Gate entrance.

By Train Take the Metro-North Harlem local line to Botanical Garden station, then cross Southern Boulevard to the garden's Mosholu Gate entrance—which is not the main entrance at Conservatory Gate.

INFORMATION

New York Botanical Garden, 2900 Southern Boulevard, Bronx, NY 10458.
General information: 718-817-8700
Directions: 718-817-8779
http://www.nybg.org, where you can also find a bird checklist and schedule of birdwalks

Binoculars may be borrowed at the visitor center's information booth. You are not allowed to use sound devices to attract birds.

HOW TO GET TO THE BRONX ZOO

By Subway Take the ❷ or ❺ to Pelham Parkway. Head west to the zoo's Bronx River entrance (Gate B) by walking south on White Plains Road across Pelham Parkway South to Lydig Avenue. Turn right and walk to Bronx Park East. Cross the street and enter the parkland on the path (playground will be on your right). Bear right at the fork in the path, then cross Boston Road and take a left onto the sidewalk. Continue through the underpass to the zoo entrance.

By Train Take the Metro-North Harlem Line to Fordham station, then travel east on Fordham Road. Or take the Bx12, Bx22 from the station along Fordham Road to the same intersection. Walk south on White Plains Road across Pelham Parkway South to Lydig Avenue. Turn right and walk to Bronx Park East. Cross the street and enter the parkland on the path (playground will be on your right). Bear right at the fork in the path, then cross Boston Road and take a left onto the sidewalk. Continue through the underpass to the zoo entrance.

By Bus From Manhattan, the BxM11 express bus makes stops along Madison Avenue between 26th and 99th Streets, then travels directly to the zoo's Bronx River entrance (Gate B). From the Bronx, take the Bx9 or Bx19 to 183rd Street and Southern Boulevard, which is the location of the zoo's Southern Boulevard pedestrian entrance (Gate C). Or take the Bx12 or Bx22 to Fordham Road and Southern Boulevard, then walk five blocks south on Southern Boulevard to 183rd Street. From the Southern Boulevard entrance, make your way through the zoo to the Bronx River Gate. From Queens, take the Q44 to 180th Street and Boston Road. Walk north one block to the Bronx Zoo's Asia gate entrance (Gate A). You can pick up Jungle World Road here, and, once you reach the Bronx River Gate, the Mitsubishi River Walk.

By Car The WCS official address, 2300 Southern Boulevard, will take you to the Southern Boulevard entrance. For the Bronx River Gate entrance, set your GPS for the intersection of Bronx River Parkway and Boston Road, Bronx, New York 10460.

INFORMATION

The Wildlife Conservation Society Bronx Zoo: 718-220-5100
http://bronxzoo.com/
Mitsubishi Riverwalk: http://bronxriver.org

Both parks charge an admission on most days. Consult their respective websites for details.

Restrooms, café, and other food services are available at both locations.

The bookstore and gift shop at the Botanical Garden has a stellar selection of books on a range of natural subjects.

Scope Not necessary.

Photos Easy.

Van Cortland Park

This large park in the northwest Bronx comprising 1,146 acres offers excellent birding in a variety of different habitats that include native woodlands, a freshwater lake, grassy areas, and wetlands. It's one of the best birding locations in the Bronx, and because of its importance during migration and breeding, as a critical habitat of forest and wetlands in an area surrounded by urbanization, Van Cortlandt has been designated an Important Bird Area. Nearly every species of local land bird has been seen in its northeast section. Van Cortlandt Park is divided into several parts by three major highways and two golf courses, so it's best to think of it as a collection of separate parks. The northeast section, made up of Croton Woods and the Northeast Forest, is a hot spot for migratory and nesting birds but is difficult to get

Northern Waterthrush

to via public transportation. It also has rather lonely and seldom-traveled portions that can seem either refreshingly unmanicured or spooky, depending on your taste. Recent converts to Van Cortlandt normally stick to the western side, where they can use the 242nd Street subway stop from the ❶ to enter at the southeast corner and, with quite a bit of walking, visit all the birding hot spots.

VIEWING SPOTS

Enter via the southwest side on Broadway at the 242nd Street entrance. Head north and west toward the *Parade Ground*. The Nature Center and restrooms are in the low building near Van Cortlandt Mansion. Scan the Parade Ground for grassland species. At the edge of the woods, pick up the *John Kiernan Nature Trail* and explore the wetlands. Pick up the *Putnam Trail* and bird along the side of *Van Cortlandt Lake* and *Tibbetts Brook*. The Putnam Trail will take you all the way to the top of the park, where you will then double back along the same route. Another idea would be to bird as far as the Mosholu Parkway Crossing and double back from there to the bridge across Tibbetts Brook taking you west. Once back at the Parade Ground, turn

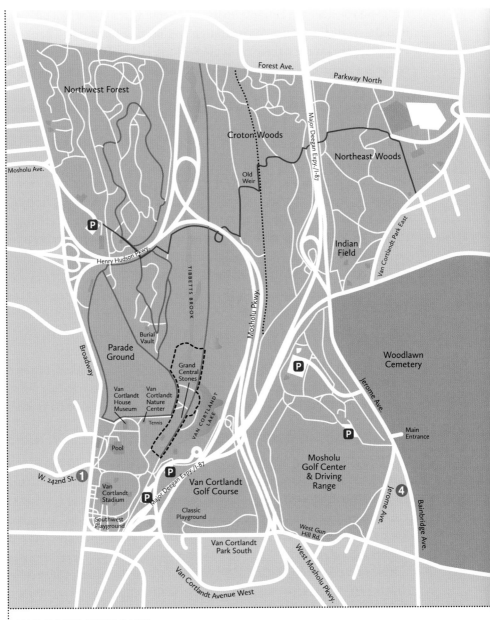

VAN CORTLANDT PARK

right and you'll see the entrance to the cross-county trail. That will take you into the *Northwest Forest*, for migrating and nesting songbirds. Thanks to cross-country runners, these trails are well-marked and easy to use. People out for a walk in the woods also frequent them and appreciate the wide paths, ease of use, and the fact that poison ivy is kept at bay.

For the more adventurous, use the Putnam Trail described above to enter *Croton Woods* and the *Northeast Forest* or enter the park at the Woodlawn Playground. From there, you can use any of the official and unofficial trails to wander around.

Please note that while the John Muir Nature Trail takes you through some pretty scenery, it also takes you right beside cars rushing along the freeway. Furthermore, it poses the temptation of taking shortcuts across the golf courses, which may lead to awkward encounters with golfers. For that reason, we do not recommend using it to move from one part of the park to another, but prefer the main north–south route, the Putnam Trail.

If you have time after making the rounds in Van Cortlandt Park, directly across Jerome Avenue on the east side of the park is lovely *Woodlawn Cemetery*—see our description in Other Places to Find Birds in the Bronx.

KEY SPECIES BY SEASON

Spring Over a dozen species of warbler, including Northern Parula, Palm, and Pine; migrating Killdeer; Northern Rough-winged, Tree, and Barn Swallows.

Summer Wood Ducks are seen year-round, but summer into fall is the best time for viewing; waders include year-round Great Blue Heron; Great Egret, and Green Heron spring through fall; Eastern Kingbird, Warbling Vireo; nesting Yellow Warbler and Baltimore Oriole.

Fall Migrating Ruby-throated Hummingbirds; Savannah and later Fox Sparrow; Hermit Thrush; American Goldfinch (which also stay through the winter).

Winter Waterfowl include Ruddy, Wood, and American Black Ducks, plus Hooded Mergansers; occasionally, the Parade Ground will have uncommon geese such as Greater White-fronted, Snow, or Cackling, but the sheer number of Canada Geese is impressive. Year-round residents include Downy, Red-bellied, and Hairy Woodpeckers; Swamp Sparrow.

HABITAT

Native woodlands, freshwater lake and wetlands, grassy areas.

BEST TIME TO GO

Spring through fall.

HOW TO GET TO VAN CORTLANDT PARK

By Subway Take the ❶ to the 242nd Street stop.
By Bus The park is accessible via the Bx9, BxM3, Bx16, Bx31, and Bx34.

HOW TO GET TO CROTON WOODS AND THE NORTHEAST FOREST

By Subway and Bus Take the ❹ to Woodlawn and the Bx34 or BxM4 up
to Van Cortlandt Park East to enter the park at the Woodlawn Playground.

By Car From the Major Deegan Expressway use the Van Cortlandt Park
South exit for the south and west portions of the park, or East 233rd Street
for the north and east. The Henry Hudson Parkway's Broadway exits will
also work.

Parking lots are located at Shandler Ballfields and Van Cortlandt Golf
House.

INFORMATION

General inquiries: 718-430-1890
Urban Park Rangers: 718-548-0912
Friends of Van Cortlandt Park: 718-601-1460
Van Cortlandt Nature Center: 718-548-0912
http://www.nycgovparks.org
Van Cortlandt Park Conservancy will provide information and a map:
   http://www.vcpark.org

There are restrooms on Broadway near 242nd Street, at the Nature Center,
and in the Northwest Forest. No restaurants, but there are a number of places
along Broadway where you can buy food and take it to the benches and many
shady areas to relax and picnic.

Because some of the trails in the northeast section are pretty wild, you can expect poison ivy encroaching on the path, as well as significant numbers of mosquitoes from time to time. Long sleeves and long pants are a good choice.

Some of the areas are not well visited, and for safety reasons it would be smart to have someone with you.

OTHER THAN BIRDING

There are lots of things to do and see apart from birding and golf: visit the Van Cortlandt House Museum, Van Cortlandt Nature Center, watch cricket matches, run a cross-country course. Van Cortlandt also has a multiuse stadium, a public pool, an equestrian center, and a seasonal skating rink.

Scope Not needed.

Photos Challenging when the trees are leafed out.

OTHER PLACES TO FIND BIRDS IN THE BRONX

Not to be overlooked are some smaller Bronx parks that can be both interesting and productive. We marked the best with an asterisk.

Woodlawn Cemetery*

Established in 1863, this beautiful and tranquil National Historic Landmark is a fine place spring through fall to see migrating and nesting birds. Often overlooked by birders, it has four hundred acres of venerable trees, open grassy areas, and a freshwater lake. As you stroll beneath the arbors and among the many elaborate mausoleums, you can search for the final resting places of Herman Melville, Irving Berlin, Miles Davis, Duke Ellington, LeRoy Neiman, and other well-known personalities while finding kinglets, warblers, Hermit Thrush, Great Blue Heron, and Wood Duck. Expect to see year-round Red-tailed and Cooper's Hawks, woodland birds, and the resident Wild Turkeys. This is a safe place to bird alone, as there is good security, but getting there without a car is best done with a pal. Take the ❷ or ❺ to 233rd Street; the ❹ to Woodlawn; the Metro-North Harlem Line local train (North White Plains) to Woodlawn station; or be prepared for a pricey cab ride. The main entrance is at Webster Avenue and East 233rd Street, Bronx, New York 10470; phone is 718-920-0500 or 877-496-6352.

Wild Turkey at Woodlawn Cemetery

Wave Hill

For something different, or to please a friend who is really into horticulture, visit this lovely nineteenth-century Riverdale estate turned public garden, which has spectacular views of the Hudson River and the New Jersey Palisades. Situated on a bank overlooking the Hudson River, Wave Hill can be a decent hawk-watching spot, and during fall migration you are at a height to also see other migrants streaming past. Get to the lawn by the river for the best unobstructed views, and let a northwest wind bring the birds to you. In winter, especially in February when the Hudson freezes, Bald Eagles can be seen—sometimes riding on the ice.

There is a fee to enter and one to park, but it is simple to get there by public transportation. Take the ❶ to 242nd Street, or you can take Metro-North to the Riverdale station. There is a free Wave Hill shuttle, which meets some of the trains at both stations. Check their website for more info on opening hours and transportation: https://www.wavehill.org. West 249th Street, Bronx, NY 10471; 718-549-3200.

Riverdale Park / Raoul Wallenberg Forest Preserve

Just south of Wave Hill is Riverdale Park, 112 acres that includes wetlands to the north. However, the southern portion has the most native plants and trees and is usually a better bird attractor. If you continue walking south through Riverdale Park, the Raoul Wallenberg Forest is located across Palisade Avenue starting around 236th Street. This is a Forever Wild forest and has great trees but is not generally frequented by birders. If you're in the area, it's worth a quick look, as there can be warblers during migration.

Spuyten Duyvil Shorefront Park

Get a different view of the "Spouting Devil" (Spuyten Duyvil to the Dutch) and its flowing creek, which are most often seen from Inwood Hill Park. This small park under the Henry Hudson Bridge and south of Riverdale Park has sweeping views of Inwood Hill Park (see our complete description of this hot spot in the Manhattan chapter). Edsall Avenue circumnavigates the park and will give you a lower and more intimate view of the area than Palisade Avenue. Check out the little pond under the bridge off Edsall by the Hudson for waterfowl, and as you walk, keep an eye on the trees for songbirds that stop off here during migration. As you would at Inwood, look for raptors over the water. The Spuyten Duyvil Metro-North railroad stop is right under the bridge and gives easy access to the park. Otherwise drive and park on the street. The ❶ has a stop at 231st Street, but it's a bit of a hike. Bird this park with a buddy.

North Brother and South Brother Islands

Both of these now-uninhabited islands in the East River between the Bronx and Rikers Island are Important Bird Areas in the Harbor Herons Wildlife Complex because of their importance as nesting sites for wading birds including Great Egret and Black and Yellow-crowned Night-Herons. From a boat, the decaying remains of a nineteenth-century smallpox hospital on North Brother Island can be easily seen. Both islands are now owned by the New York City Parks Department and are off-limits to visitors. Some boat excursions will take you past them, and during nesting season these active rookeries can be seen from the water (see "The Best," New York Harbor Boat Tours).

Staten Island

Characterized by a surprising diversity of habitats and birding opportunities, Staten Island often doesn't get the attention it deserves. The southernmost of the five boroughs, and the least populated, Staten Island got its name in the seventeenth century from the Dutch, who called it Staaten Eylandt (States Island). It's reachable only by car, express bus, or ferry, but if you make the effort to visit, you'll find a remarkable variety of beautiful, bird-rich parks, many of which are quite wild and some of which are found in the Greenbelt and Bluebelt systems. Local birders are friendly and willing to help.

Clove Lakes has a bird list that rivals Central Park and is the centerpiece of Staten Island birding. Although this premier hot spot is inland, a lot of action also takes place along the coastline. *Mount Loretto Unique Area*, *Blue Heron Park*, and *Conference House Park* at Staten Island's southernmost tip complete the list of top destinations. While Hurricane Sandy damaged *Great Kills* and *Wolfe's Pond*, they are still worth visiting. If you spend some time exploring, you will also find remote swamps like *Clay Pit Ponds Preserve*, wild-managed sliver parks wedged between residential areas, productive wetlands behind shopping malls, one of only a handful of Purple Martin nesting colonies in New York City at *Lemon Creek*, Staten Island's hawk-watching hill *Moses Mountain*, dense forests, and an authentic Chinese Scholar's Garden at *Snug Harbor*.

To bird Staten Island efficiently, moving around by car is best. However, if you take public transportation, don't despair—there is a rail and bus system that will get you to many of the places we describe. The Staten Island Railway runs the full length of the island from the ferry terminal in St. George to Tottenville, and unlike the bus, for one fare it offers hop-on and hop-off service. If you are coming from Manhattan, enjoy the venerable Staten Island Ferry. It is free and gives a stellar thirty-minute trip across New York Harbor, past the Statue of Liberty and Governors Island.

Newark Liberty
International Airport

81

NEWARK BAY

BAYONNE

440

UPPER NEW YORK BAY

Shooters
Island

Bayonne
Bridge

Snug Harbor

ST. GEORGE

Ferry
Terminal

BROOKLYN

278

Bridge
Creek

Mariner's Marsh

Richmond Terrace

Allison Pond

Goethals
Bridge

Goethals Pond

Silver
Lake

NEW JERSEY

1

Old
Place

Clove
Lakes

Verrazano
Narrows
Bridge

95

278

Prall's
Island

South Ave.

Staten Island Expressway

Ft. Wadsworth

440

Willowbrook
Park

MOSES
MTN.

High
Rock
Park

New Jersey Turnpike

Fresh
Kills

South
Beach

Isle of
Meadows

Latourette
Park

New
Dorp

Midland
Beach

Greenbelt

Richmond Ave.

King
Fisher
Park

Miller
Field

ARTHUR KILL

STATEN ISLAND

Hylan Blvd.

LOWER NEW YORK BAY

Clay Pit
Ponds

Great
Kills
Park

Oakwood
Beach

440

Long
Pond

Blue
Heron
Park
Preserve

Tottenville
Train Station

Lemon
Creek

Wolfe's
Pond Park
and Acme
Pond

TOTTENVILLE

Mt.
Loretto

Conference
House

Sandy
Hook

RARITAN BAY

STATEN ISLAND

Clove Lakes Park

For the best place to see warblers on Staten Island during migration, and one of its birding highlights, head straight for Clove Lakes. This easy-to-access and lovely Forever Wild park is a well-known warbler trap that draws in birders as well. People who use this park are happy to direct you to the birds. With nearly one hundred acres of wild woodlands surrounding several freshwater lakes, this is a nice spot to visit nearly year-round, but migration is when this place shines. At peak times, you may be able to find up to seventy-five species in one day. No need to worry about getting lost here—the paths are well marked, and oftentimes an easy walk around the lakes and along the streams will net some great birds. While you are visiting these pretty lakes, look for a three-hundred-year-old tulip tree that is said to be the oldest living thing on Staten Island.

VIEWING SPOTS

Clove Lakes Park comprises woodlands and three freshwater lakes. From north to south they are Brooks, Martling, and Clove Lakes. There are a number of ways to tackle this park, but the following are the prime warbler territory and hot spots not to be missed during migration:

the stream that runs between Martling Lake and Clove Lake

the stream that runs between Martling Lake and Brooks Pond

the overgrown woods at the north end of the grassy field at the Royal Oak entrance to the Park

the mid-level bridle paths that parallel Martling Lake and the stream connecting it to Clove Lake

There are a number of places to enter the park. To focus on the prime areas, try this suggested route: Park on the street on Royal Oak Road and enter off Royal Oak where it turns into Rice Avenue. At this entrance, right in front of you is an open grassy field. To your left the field borders an overgrown wood where the *Big Sit* is held every October. Birders gather all day in this one spot tallying as many birds as they can find. Sixty-five birds have been counted

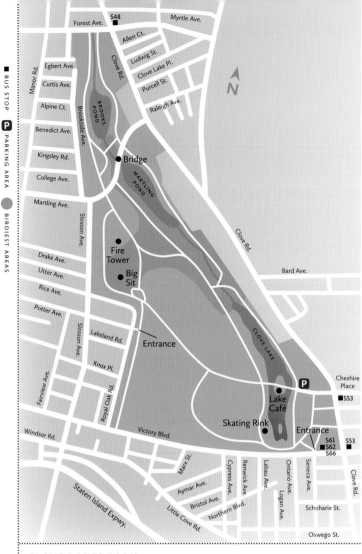

CLOVE LAKES PARK

here in a single day. Walk along the wooded area on the north border of the field, scanning the skies for high-flying migrants in spring and fall, and check out *the fire communications tower* for Red-tailed Hawks.

Continuing along this path, when you see the sign for "Clove Lakes 5K Trail," take that to the right. This is the higher of the two trails and winds past the lower end of *Martling Lake*. When you can, transfer to the lower path, and you will find yourself in the warbler hot spot on the *stream between Martling and Clove Lakes*. If you want to take advantage of two hot spots on this route, cross over to the east side below Martling Lake and take the trail along the stream to Clove Lake, paying particular attention during migration to the brushy areas alongside the path close to the stream. Stay on the low path following the stream, keeping an eye open for Scarlet and Summer Tanagers, flycatchers, and vireos.

When you reach the bridge at the top of *Clove Lake*, you can make a choice. If you are feeling energetic, you can continue in the same direction and stroll around Clove Lake. The walk around is less birdy but scenic, and you can look along the lakeshore for wading birds. In winter there are waterfowl. Try to get there before midday, as even in winter this is a popular park with enough jogging, dog walking, and other activity to send some of the less fearless species elsewhere once the crowd arrives.

This lake is less of a bird magnet than the *area between Clove and Martling Lakes*, which can be one of the birdiest places on Staten Island during migration. So if you want to continue the hot spot route, turn around and take the path on the other side of the stream back toward Martling Lake. Here you can focus on the *mid-level bridle path* on the way. Once you get to *Martling Lake*, take the low path to Martling Avenue, where you may be able to see the Great Blue Heron nest in a tree near the bridge. Along the way also look for ducks, egrets, and cormorants.

From this point, if you have time, cross Martling Avenue and follow the paths north along the stream for the last of the birdiest spots, *Brooks Pond*.

KEY SPECIES BY SEASON

Spring Songbirds including over thirty species of warbler, six different species of vireo, Baltimore Oriole, Blue-gray Gnatcatcher. Nesting Great Blue Herons are easily seen; Woodcocks have been reported; Belted Kingfisher, Eastern Phoebe, Great Crested Flycatcher, Scarlet Tanager.

Summer Nesting Great Blue Heron, Great Egret, Black-crowned Night-Heron; Spotted and Solitary Sandpipers; Osprey, Ruby-throated Hummingbird, Scarlet and Summer Tanagers, Eastern Wood-Pewee and a variety of flycatchers.

Fall Migrating waterfowl, songbirds returning including warblers, Purple Finch; Chimney Swift, Northern Flicker, Cedar Waxing; a variety of sparrows including Chipping, Field, Savannah, Song, White-crowned, Swamp; Pine Siskin; Sharp-shinned, Cooper's, and Red-tailed Hawks; Winter, House, and Carolina Wrens.

Black-throated Blue Warbler

Winter Waterfowl such as Wood Duck and American Wigeon, Hooded Merganser, American Coot; Downy, Hairy, and Red-bellied Woodpeckers; White-throated Sparrow, Red and White-breasted Nuthatch.

HABITAT

Woodlands, edge, freshwater lakes.

BEST TIME TO GO

Spring is a highlight, although other seasons also net good views and varieties of birds. Summer brings wading birds and flycatchers, plus many nesting birds. Fall migration is best mid-August through November for songbirds, and in winter there are waterfowl.

HOW TO GET TO CLOVE LAKES PARK

By Bus From the St. George Ferry Terminal take the S61, S62, or S66 to the park entrance at the intersection of Victory Boulevard and Clove Road; the S48 on the north side of the park along Forest Avenue; or the S53 on the east side on Clove Road.

By Car For this itinerary, park on the street on Royal Oak Road and enter as indicated above at Rice.

INFORMATION

http://www.nycgovparks.org/parks/CloveLakesPark
http://www.nycaudubon.org

Restrooms are in the Parks Department building near the boat rental concession. There is a restaurant on the island in the middle of Clove Lake.

This is a fairly safe park with friendly birders around, but since there are some remote spots, it's probably a good idea not to go here alone.

Watch out for ticks and poison ivy.

OTHER THAN BIRDING

This park has well-maintained woodlands and paths and lots of other public-use spaces, especially in the southern portion. They include a fitness path, basketball courts, baseball and soccer fields, and in season, ice-skating and paddle-boat rental. Fishing is permitted. Barbecue pits, picnic tables, and benches are found throughout the park.

Scope Not necessary.

Photos Easy.

Great Kills Park

Marketed as both a beach playground and a place to view wildlife, this six-hundred-acre park is part of the Gateway National Recreation Area. There are birds here year-round, but you may want to avoid the crowds and visit in fall and winter. Great Kills is situated on a long peninsula with a central road that runs past beaches, marsh, woodlands, and a marina. Side trails wander through some birdy areas. The range of habitats makes it a good spot to see a large variety of birds, but there is also a lot of human activity. Check it out in early September for huge numbers of swallows gearing up to fly south. Later in the year, you will reliably see Horned Larks, shorebirds, waterfowl, and often Snow Buntings. If you visit in summer, Great Kills also has nesting Osprey and a Purple Martin colony near the marina. Owls live here, too. The park was greatly affected by Hurricane Sandy, and even years after the storm, debris of all sorts remains part of the landscape. Because radioactive contamination was discovered in the area to the northeast of Great Kills Bay in 2005, that section of the park is closed to the public until cleanup has been completed.

VIEWING SPOTS

Most people get around Great Kills by car, but you can do it by bike, too. Buffalo Street is the main central road, and it stretches the full two miles of the peninsula—all of which you will want to visit. As you enter the park, bypass the first

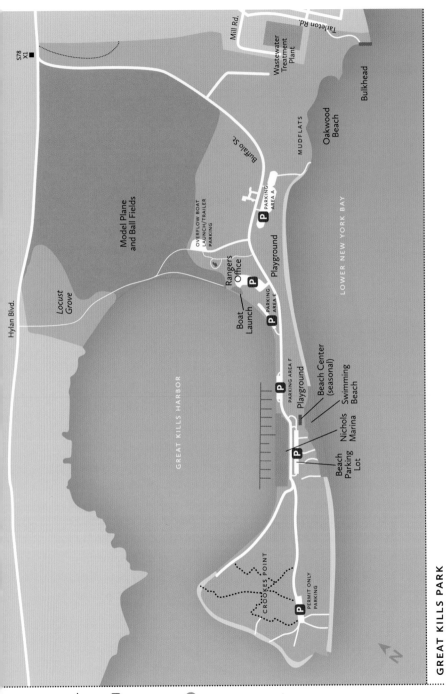

GREAT KILLS PARK

S78
X1 ■

Hylan Blvd.

Locust
Grove

Model Plane
and Ball Fields

GREAT KILLS HARBOR

Mill Rd.

Tarleton Rd.

Wastewater
Treatment
Plant

Buffalo St.

Overflow Boat
Launch/Trailer
Parking

Rangers
Office

Boat
Launch

P PARKING
AREA E

Playground

P PARKING
AREA A

MUDFLATS

Oakwood
Beach

Bulkhead

LOWER NEW YORK BAY

PARKING AREA F

P

Playground

Beach Center
(seasonal)

Swimming
Beach

Nichols
Marina

Beach
Parking
Lot

P

CROOKES POINT

PERMIT ONLY
PARKING

P

N

⋮ TRAIL   P PARKING AREA   ⬤ CLOSED

parking lot and drive to the parking lot on your left, Parking Lot A, a bit farther on so that you are near the *mudflats*. Try to time your stop for low tide, to give yourself the best chance at shorebirds. Killdeer, oystercatchers, grackles, doves, Little Blue Heron, and other wading birds are found here. You might get lucky and see a Bonaparte's Gull or Royal Tern, depending on the time of year. You can see these mudflats from another perspective at Oakwood Beach—see our description later in this chapter.

Exit this parking lot and your next stop will be the *marina*, where there are two parking lots. The first is for boat trailers, the second for cars. If you're not towing a boat, don't park in the boat trailer parking lot. A group of Purple Martin houses can be seen from here, and several varieties of swallows can be found foraging in the summer. At the marina, you'll find gulls, terns, and a walkway to

Snowy Egret on mudflats

the right that will lead past the park ranger's office. Follow this path through a *grove of locust trees* that can be particularly good in the fall for migrant sparrows, vireos, warblers, flycatchers, and buntings. Farther along, near the former *model airplane fields*, Woodcock make their spring courtship displays. Once at the end, return via the same path and drive to your next stop.

On the left side of the street is a *beach parking lot*, the last parking area that doesn't require a permit. We recommend you walk to the farthest lot *alongside the main road*, where there can be a lot of birds in the trees and scrubby edges. There are *paths to your right* that go through good bird habitat. They are worth exploring, but be careful of poison ivy! During nesting season you will find Northern Mockingbirds in great numbers, Mourning Doves, House Finches, American Goldfinches, Brown Thrashers, and Yellow Warblers. Check the playground for Horned Larks in winter.

Returning to the main road on foot, continue walking toward the end of the peninsula to the permit-required parking lot, and take a walk out onto the *beach toward Crooke's Point*. This is a good viewing spot in the winter for waterfowl such as Long-tailed Ducks, loons, grebes, and Northern Gannets. It's the remotest area of the park, and if there are owls around, this would be the best spot to find them.

If you have time on your way home or missed the tide at the mudflats, stop by *Oakwood Beach* at the foot of Tarleton Street—see our description in Other Places to Find Birds on Staten Island. Just north of Oakwood is *Miller Field*, *Midland Beach*, and *South Beach*, which are also described separately in this chapter.

KEY SPECIES BY SEASON

Spring Songbird migrants including Common Yellowthroat, Eastern Towhee, Warbling and White-eyed Vireos, American Woodcock, Bonaparte's Gull, Ruddy Turnstone, Greater Yellowlegs, Spotted and Least Sandpiper.

Summer Osprey, Ring-necked Pheasant, Great and Snowy Egrets, Little Blue Heron, Black-crowned Night-Heron, Glossy Ibis, Killdeer, American Oystercatcher, Black Skimmer, Laughing Gull; Least, Common, and Forster's Terns; Willow Flycatcher, Purple Martin; Northern Rough-winged, Tree, Bank, and Barn Swallows; Yellow Warbler, Brown Thrasher, Baltimore Oriole, Field Sparrow.

Fall Common Loon, migrating shorebirds including Black-bellied Plover, Dunlin, Sanderling; possibly Royal Tern; great for raptors and is reliable for Red-shouldered and often Rough-legged Hawks; American Kestrel, Merlin, Peregrine Falcon, Cooper's Hawk; Chipping, Field, and Savannah Sparrows; Ruby-crowned Kinglet, Eastern Phoebe.

Winter Waterfowl including scaup, Long-tailed Duck, Common Goldeneye, Red-breasted Merganser, Bufflehead, Red-throated and Common Loons, Horned and Red-necked Grebes; Northern Gannet, Great Cormorant, Northern Harrier, Purple Sandpiper, Horned Lark, Snow Bunting, Lapland Longspur; American Tree, Fox, and White-crowned Sparrows. Snowy Owls appear, especially during irruption. Rough-legged Hawks have been seen.

HABITAT

Mudflats, beach, woodlands, edges, freshwater pond.

BEST TIME TO GO

Year-round, but especially good in fall and winter for interesting birds and fewer humans.

By Bus The S78 and X1 stop on Hylan Boulevard. Unless you've brought your bike, be prepared for a long walk to the end of the peninsula.

By Car The entrance is off Hylan Boulevard at Buffalo Street.

INFORMATION

3270 Hylan Boulevard, Staten Island, NY

http://www.nyharborparks.org

http://www.nycgovparks.org

http://www.nps.gov/gate

There are restrooms at most of the parking areas. The ones at the main beach parking lot are by far the nicest.

Scope A good idea.

Photos Easy.

Blue Heron Park Preserve

Sculpted iron gates mark the entrance to this beautiful 250-acre forest-and-ponds park on Staten Island's south shore. It took a handful of local residents over thirty years to piece together this preserve from several parcels of land, and it's been worth the effort. This former dumping ground for old cars and trash has been restored, and its usefulness to wildlife has been sufficiently reestablished that it is now a Forever Wild park. There are a half-dozen ponds, plus freshwater marsh, meadows, streams, and woodlands you can explore along a few well-marked paths that extend as far as Raritan Bay. This park is lovely to walk through and is a good spot to see a variety of birds. In summer and fall it's also a surprisingly good place for butterflies.

VIEWING SPOTS

The park is divided by Poillon Avenue. On the western side, behind the Nature Center and parking lot, are a couple of pretty woodland trails. Follow the *Spring Pond Loop Trail* to see Blue Heron's largest pond and a marshy area where you can find a variety of waders. The Nature Center staff also put out feeders, which are worth a visit.

Across Poillon Avenue is the larger portion of Blue Heron Park, and here you can pick up the trails to *Butterfly Pond* and *Blue Heron Pond*, which birders

Green Heron

normally find more productive. The dense forest gives refuge to a variety of migrants and breeding songbirds, while secluded Blue Heron Pond can be attractive to waterfowl, including shy Wood Ducks.

South from Blue Heron Pond is a trail that will take you across Hylan Boulevard, through a densely forested area to Seguine Pond and then Raritan Bay. At the time of this writing, the effects from Hurricane Sandy were still very evident at Seguine Pond.

If you have more time to spend in the area, visit *Arbutus Pond*, just west of Blue Heron Park and south of Hylan. From October to March there can be a variety of waterfowl, including Wood Duck, Ring-necked Duck, Bufflehead, and Hooded Merganser.

Another larger park to include if you are nearby is *Wolfe's Pond* and *Acme Pond*. See separate description in this chapter.

KEY SPECIES BY SEASON

Spring Great and Little Blue Herons, Black-crowned Night-Heron; possibility of American Woodcock; Eastern Phoebe, Red-eyed Vireo, kinglets, Baltimore Oriole, House Finch, and at least twenty species of warbler, including Ovenbird. Eastern Screech Owls are year-round residents.

Summer Herons, including Green and Great Blue; Glossy Ibis may be seen; Spotted Sandpiper, Belted Kingfisher, Great Crested Flycatcher, Eastern Kingbird, Wood Thrush, Common Yellowthroat, Eastern Towhee, and breeding backyard birds.

Fall Raptors, including Northern Harrier, Sharp-shinned and Cooper's Hawks; several species of woodpecker.

Winter Waterfowl such as Wood Duck, Gadwall, Bufflehead, and Red-breasted Merganser; Sharp-shinned, Red-shouldered, and Red-tailed Hawks; Red-bellied, Downy, and Hairy Woodpeckers, as well as Northern Flicker; Dark-eyed Junco, and White-throated Sparrow.

HABITAT

Woodlands, meadow, marsh, freshwater ponds.

BEST TIME TO GO

Early spring through summer.

HOW TO GET TO BLUE HERON PARK

The park is in the village of Annadale between Hylan Boulevard and Amboy Road.

By Bus Take the S78 or S59 to Poillon Avenue.

By Staten Island Railway Exit at Annadale station.

INFORMATION

222 Poillon Avenue, Staten Island, NY
http://www.nycgovparks.org
http://www.preserve2.org/blueheron

The trails are open every day. The Nature Center has a more limited schedule but offers numerous programs and walks and has clean bathrooms. The picnic area is handy to the parking lot.

Scope Not useful here.

Photos Easy.

Wolfe's Pond Park and Acme Pond

This is one of Staten Island's largest parks, and it has something for everyone. With popular beaches and mountain biking trails, birders might find it a bit busy. But there are pretty freshwater ponds and natural areas with hiking trails in this three-hundred-plus acres of diverse habitat. Birds can be found throughout. The primary birding attraction at this park, Wolfe's Pond, was breached during Hurricane Sandy and is no longer the productive freshwater bird haven it once was. Birds can still be found here, but it's a different group of species and, frankly, not as interesting. If you want to see the park at its best, visit the beach in the winter. In other seasons, walk through the beautiful wooded section on the other side of Hylan Boulevard and around Acme Pond and the other less-visited glacial ponds in North Wolfe's Pond Park.

VIEWING SPOTS

The *parking lot* can be a good place to start when there are birds along the edges and in front of the lot around the picnic tables. From here, walk toward *Wolfe's Pond* to look for wading birds, ducks, and geese. Damage from Hurricanes Irene and Sandy have taken their toll here and changed the character and salinity of the pond. Formerly a frequently visited freshwater pond, it is no longer the bird attractor that it once was. In summer, the water level can get very low, and you may see more mud than water. Casual visitors enjoy feeding the Mallard duck families. Occasionally, you'll find a heron or an egret.

After checking out the pond, walk back toward the *beach* and scan for shorebirds, gulls, and waterfowl along the shoreline. In winter, there can be some unusual birds hanging out just offshore. Iceland Gulls often overwinter on the beach. Excellent winter waterfowl viewing can be had at the *beach overlook*. Expect to see loons all winter and ducks and shorebirds in the spring and fall.

Sanderling

Acme Pond, which is part of North Wolfe's Pond Park, is just across Hylan Boulevard and should be your next stop. This wooded area has well-marked paths and some boggy areas around the *two freshwater ponds*. This can be a quiet location to look for kingfishers, Red-winged Blackbirds, waders, and hunting swallows. Turtles abound, and the occasional bullfrog will sound off in summer. Vernal pools appear in season, and since the area can be very wet, prepare for mosquitoes and, as always, poison ivy.

Be aware that the woods at Acme Pond are also used by mountain bikers, especially on weekends.

KEY SPECIES BY SEASON

Spring Migrating songbirds, including over a dozen species of warbler, some shorebirds; Black-crowned Night-Heron, Bonaparte's Gull; Orchard Oriole.

Summer Expect to see baby ducks and Great and Snowy Egrets; Eastern Kingbird, Warbling Vireo, Baltimore Oriole; Barn, Bank, and Tree Swallows, as well as Purple Martins; shorebirds including Spotted and Least Sandpipers and American Oystercatcher at the beach; kingfishers, Great Egret, Red-winged Blackbirds, woodpeckers, goldfinch, catbird, mockingbird, and other common birds at Acme Pond.

Fall Osprey, Great Blue Heron, Forster's Tern; Ruddy Turnstones, Black-bellied Plovers, and White-rumped Sandpipers migrate through; Savannah, Song, and White-throated Sparrows; Downy, Hairy, and Red-bellied Woodpeckers.

Winter Iceland Gulls overwinter here. Expect to see Red-throated and Common Loons all winter. The waters offshore have other interesting winter waterfowl, including Red-breasted Merganser, Horned Grebe, Long-tailed Duck, and Common Goldeneye.

HABITAT

Beach, marine, woods, and some marsh and small freshwater ponds.

BEST TIME TO GO

Winter for waterfowl; summer and fall for shorebirds and migrants.

HOW TO GET TO WOLFE'S POND PARK

By Bus The park is served by the S55, S56, S59, and S78.

By Staten Island Railway Get off at the Prince's Bay stop, head South on

Seguine Avenue, then left on Herbert Avenue and into the park. This will start you in the northern, wooded area and on a hike to Acme Pond. To go directly to the beach, use Seguine Avenue to reach Hylan Boulevard, turn left, and look for the trail on the right just past Wolfe's Creek Park.

By Car Driving is the best way to reach the park and its two main areas. The beach parking lot is located off Hylan Boulevard; turn south on Cornelia Avenue and right at the T intersection onto Chester Avenue. For Acme Pond and its surroundings, you can either cross Hylan on foot or take the car and park it on Chisholm in front of the trailhead (opposite the loading dock at the high school).

INFORMATION

http://www.nycgovparks.org
http://www.nycmtb.com/?page_id=289

There is a very nice, modern restroom facility at the beach, BBQ pits, and picnic tables scattered between the parking lot and the beach.

No food service. Keep an eye out for poison ivy, and mosquitoes can be an issue on the Acme Pond side of the park.

Scope Would be helpful at the beach for waterfowl and near Acme Pond where access is limited and views across the lake may require a scope.

Photos Medium difficulty. There can be a lot of foliage, and birds are often at a distance.

Lemon Creek Park

Otherwise known as Prince's Bay, this eighty-acre park bisected by Hylan Boulevard has two distinct features. The pier that juts into Raritan Bay can be productive year-round, and the park itself has one of the only Purple Martin colonies in New York City. Lemon Creek also has woodlands and the Lemon Creek Tidal Wetlands—which are not as easy to access. While not a top birding location, this park still has its fans, and you may decide it's worth a stop when you are in the area.

VIEWING SPOTS

The pier offers nice views of overwintering waterfowl and the gulls that loiter in the area year-round. If you have a scope, bring it. Check out the edges along

Purple Martins

the shore for migrating birds in season and the odd Bonaparte's Gull. There may be Osprey overhead, occasional Black Skimmers working the shallows, and shorebirds poking around along the shoreline in summer. Bald Eagles are also seen from this location. In winter look for common ducks along with Long-tailed Duck, scaup, Common Goldeneye, Red-breasted Merganser, Red-throated and Common Loons, Horned and Red-necked Grebes. In spring and summer the scrubby area behind the pier may net some warblers and migrating songbirds. Use the parking lot where Sharrott Avenue dead-ends into the beach.

The marina is the place to see Purple Martins April through August. Take Seguine Avenue toward the water and park in the lot, walk back toward Johnston Terrace, and make a left there. The colony is easy to spot.

As you come back to the car, look for birds in the trees and bushes surrounding the parking area. If you want a short beach walk, there are a few paths giving different views, but this is not the most pristine coastline.

HOW TO GET TO LEMON CREEK PARK

The pier is located at the end of Sharrott Avenue. Lemon Creek Park is just west of Wolfe's Pond Park, south of Hylan Boulevard. Parking is limited and is accessed via Seguine Avenue.

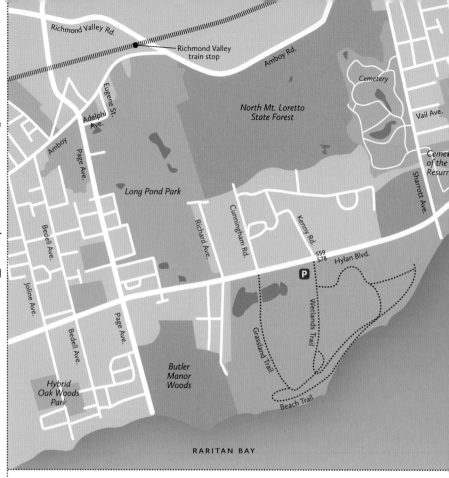

MT. LORETTO UNIQUE AREA

Mount Loretto Unique Area and
North Mount Loretto State Forest

This diverse park on Raritan Bay is one of the highlights of Staten Island birding year-round and a favorite with locals. With five habitats in its two hundred acres of productive grassland and wetlands, as well as nearly one hundred acres of forest, there are lots of opportunities to see a variety of birds. In the late 1800s through the 1970s Mount Loretto was an orphanage larger than Boys Town. Now it is an attractive respite for grassland and marsh-loving birds and wildlife, with attractive woodlands and scenic views from its shoreline.

Mount Loretto comprises two separate areas. The *Unique Area* is the larger of the two and has stunning views over Raritan Bay. While mostly grassland, it actually has a variety of habitats and will be the most productive part of your trip. The *State Forest* across Hylan Boulevard will keep you busy with woodland birds.

There are three trails in the Unique Area—Grasslands, Beach (which has an amazing overlook to the bay), and Wetlands—all of them worth visiting. You really cannot go wrong here following any of the trails and will be rewarded by interesting birds in each habitat.

In addition to sparrows, the fields along the *Grasslands Trail* are full of wildflowers and butterflies in spring and summer. Fall, when the fields are golden, can provide a transporting experience. Make sure you look overheard for raptors—many of which like cruising the grasslands as well. The fields in the Unique Area are mown only once per year before Thanksgiving to give grassland birds the most refuge and nesting opportunities.

In summer, take the *Beach Loop Trail* and look for Bank Swallows nesting in the banks under the lighthouse. In winter, the seventy-five-foot-high *bluffs* over the bay offer opportunities to see Harbor Seals hauled out onto the rocks and waterfowl on the open water.

Indigo Bunting

Apart from the trails, especially during migration, check the *thickets* on the woodland side of Hylan Boulevard for migrating warblers. Bald Eagles and other raptors can also be seen from Hylan.

If you have the time, also visit the *North Mount Loretto State Forest* across Hylan Boulevard. These ninety-four acres are made up of nearly half wetlands and half upland forest. While pretty, it's not quite the hot spot that the Unique Area is, so if your time is limited, choose the Unique Area over the State Forest. You can leave your car in the parking lot and walk to the forest, or you can drive and park on the side of Amboy Road to enter, but this is a remote location, so be sure to go with someone.

Contiguous with this forest is *Long Pond Park*—see our description later in this chapter.

KEY SPECIES BY SEASON

Spring Waterfowl persist; expect to see a variety of songbirds that prefer open country, such as Indigo Bunting, Killdeer, sparrows, and swallows; good time to look for Bald Eagles.

Summer Nesting species include Wood Duck, Orchard Oriole, Yellow-billed Cuckoo, White-eyed and Warbling Vireo, Willow Flycatcher, Cedar Waxwing, Indigo Bunting, American Goldfinch, Common Yellowthroat. You will also see egrets, herons, Osprey, terns, flycatchers, and swallows, including nesting Bank Swallows.

Fall Bluebirds; a variety of migrating raptors; Golden and Ruby-crowned Kinglets; Flickers and woodpeckers; Red-headed Woodpeckers have been seen in the woods; Sparrows and possibly Bobolinks.

Winter In addition to common overwintering birds, American Tree Sparrows, wrens, and waterfowl. Rarely do people find Short-eared and Long-eared Owls, but Great Horned and Eastern Screech Owls are year-round residents. From the bluffs see Great Cormorant, loons, grebes, Greater Scaup, Long-tailed Ducks, and Common Goldeneye among other waterfowl; Northern Harriers and Red-shouldered Hawks.

HABITAT

Marine/coastal, grassland, forest, and tidal and freshwater wetlands.

BEST TIME TO GO

Year-round, but spring through fall is when this area is in its prime.

By Bus The S59 and S78 stop across from the entrance to the Unique Area, on Hylan Boulevard.

By Car The Mount Loretto Unique Area parking lot is located south of Hylan Boulevard between Sharrott Avenue and Page Avenue.

INFORMATION

6450 Hylan Boulevard, Staten Island, NY
DEC Region 2 Office: 718-482-4942
Forest rangers: 718-317-8213
http://www.dec.ny.gov/outdoor and search "Loretto"

There is a portable toilet near the parking lot off Hylan Boulevard. No café or food services.

For your own safety, stay on the trails. Summer brings poison ivy, ticks, and mosquitoes, so be prepared.

Scope Would be helpful for waterfowl.

Photos Easy.

Long Pond Park

Contiguous with North Mount Loretto State Forest, this park contains 115 acres of Forever Wild freshwater wetlands, forest, and grassland. It includes seven ponds, with spring-fed Long Pond as the largest. This park is productive spring through fall, with some interesting nesting birds in summer.

VIEWING SPOTS

Enter at Page and Adelphi Avenues and take any of the trails to explore the woodlands and ponds—shown on the map of Mount Loretto. Look for Wood Ducks and shorebirds on Long Pond, and in spring try for warblers and songbirds. The woods support nesting birds like Red-eyed Vireo and Eastern Towhee. No need for a scope here; dress in long sleeves and long pants, as there are lots of ticks.

HOW TO GET TO LONG POND PARK

By Bus Take the X17, X22, or X22A to the Amboy Road and Page Avenue stop.

Conference House Park

Named for a seventeenth-century manor house where an unsuccessful peace conference took place during the Revolutionary War, Conference House is part of a 265-acre park whose lengthy expanse on open water and sweeping views of Raritan Bay make it a great spot in winter for waterfowl. The rest of the year this highly recommended spot brings in songbirds (especially sparrows), shorebirds, and birds that use wetlands.

VIEWING SPOTS

Enter the park on Satterlee Street and look for the historic manor house on your right. In spring and fall, you will want to visit the *ancient oak tree* just past the house and check the tangles in the woodlands behind the tree for migrants. You may find Indigo Buntings and Bluebirds as well. Take your pick of paths through the woods, or head toward the *beach*, where in winter you can set your scope and look for loons and ducks in Raritan Bay. Keep an eye out for the red South Pole, a good orientation landmark that notes the southernmost point in New York State.

At the beach, Northern Gannets may be seen. Scan the shoreline for Bald Eagles that sometimes perch here.

After walking on the beach, go inland at the South Pole toward the *woodlands* in summer and during migration. Follow the trail to the attractive *wetlands* area in the east and look for Red-winged Blackbirds and waders. This part of Conference House Park is also home to muskrats. If you walk around the wetlands, the path will eventually take you back to the parking lot.

KEY SPECIES BY SEASON

Spring Eastern Phoebe, kinglets; warblers including Yellow, Yellow-rumped, Palm, American Redstart, and Black-and-white; Eastern Towhee, Indigo Bunting, Bluebird.

Summer Red-winged Blackbird, Baltimore Oriole; waders include Great Egret.

Fall Sparrows, such as Savannah, White-throated, Chipping, and Eastern Towhee; Cedar Waxwing; Downy, Hairy, and Red-bellied Woodpeckers; returning warblers, Osprey.

Winter Waterfowl including large numbers of Brant, Common Goldeneye, Red-breasted Merganser, Red-throated and Common Loon; good numbers

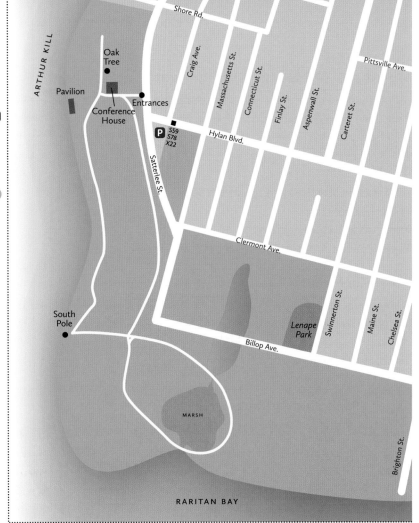

CONFERENCE HOUSE PARK

of Horned Grebe; Northern Gannet; occasionally Bald Eagle; Black-capped Chickadee.

HABITAT

Woodlands, beach, and wetlands.

BEST TIME TO GO

Year-round, but winter, fall, and spring, in that order, are best.

By Bus S59, S78, and X22 all stop at Craig and Hylan.

By Car Car is the easiest way. Follow Hylan Boulevard south to its end, then turn left onto Satterlee, which takes you into the parking lot.

INFORMATION

298 Satterlee Street, Staten Island, NY 10307

718-984-6046

http://www.nycgovparks.org

There are very nice restrooms in the visitor center; no café.

OTHER THAN BIRDING

Biking trails, kayak launch, and historic sites. The Conference House is a National Historic Landmark and New York City Landmark.

Scope Yes in winter.

Photos Long lens is helpful.

Clay Pit Ponds State Park Preserve

The only state park on Staten Island, and an Important Bird Area, Clay Pit Ponds Preserve is 260 acres of wild marshes, sand barrens, spring-fed streams, and woodlands. The Lenape Native Americans lived here. In the nineteenth century it was mined for the kaolin clay needed for the bricks used to build New York City, but after the abandonment of the mining operation, the area was not maintained. Now there are a variety of bird habitats, and for many years the park harbored grassland birds. Clay Pit Ponds can be a real hot spot during migration, with eighty birds recorded during Big Sits at Sharrotts Pond and 180 birds recorded over the years.

Times are changing in this transitional preserve. Sadly, Bobwhite no longer can be found here, and Whip-poor-will are far less common and no longer nest here. There is increasing pressure at the park boundaries for service roads, and encroachment means that the previously wild northern section is now accessible via a trail. Neighbors would like to repurpose this haven for active recreation. Mountain biking, fishing, boating, and dog walking would certainly make the preserve less attractive to birds and wildlife.

Mallard

VIEWING SPOTS

The key to Clay Pit Ponds is time. This is not a place to make a quick stop and move on. You'll want to visit and revisit, as the birding can be hit-or-miss. The best spot is *Sharrotts Pond*, in the southeast corner. Use the observation deck and have your own Big Sit. With patience, this is where you will see the highest diversity of birds.

The area above Clay Pit Road used to be inaccessible, but now there are hiking paths through it. The birdiest spot in this section is the old *gun range*, where Baltimore Oriole and White-eyed Vireo nest. It's the easiest place to get to in the northern section and is reached from Clay Pit Road—look for where the stream intersects the trail. Even though the gun range closed years ago, traces of it remain.

If you have time, you may want to check out *Abraham's Pond* and *Ellis Swamp* in the western section below Clay Pit Road, and *Tappen's Pond* in the northeast. All are interesting to hike but are secondary to Sharrotts Pond and the gun range for birds.

HOW TO GET TO CLAY PIT PONDS

The preserve is located in the southwest part of Staten Island between Arthur Kill Road and the West Shore Expressway. Park at the Interpretive Center at 83 Nielsen Avenue.

83 Nielsen Avenue, Staten Island, NY 10309

718-967-1976

http://www.nysparks.com

There are distinct hiking and bridle paths. Be sure to stay on the hiking paths—the bridle paths are for horses and riders only.

Clay Pit Ponds is heavily infested with ticks. Long sleeves and long pants are a good idea.

OTHER THAN BIRDING

The park offers a variety of nature-oriented walks and programs.

High Rock Park and Conservation Center and Moses Mountain

These beautiful ninety acres of dense rolling woodlands, tranquil freshwater ponds, and wetlands are a highlight of the Staten Island Greenbelt and serve as its headquarters. Part of the land was rescued from destruction by the never-completed Richmond Parkway, and has been maintained in its natural state. Spring Peepers still herald the start of spring in this Forever Wild–designated park, and you can take advantage of the rambling trails to productive wetlands and a wide variety of habitats that are for good birding during migration.

VIEWING SPOTS

There are two entrances to this park, but the best start for birders is the main entrance at the end of Nevada Avenue. From here you can pick up the Yellow Trail, which runs for many miles through the Greenbelt. Your primary goal at this end of High Rock is to bird the woodlands and see the wetland—*Loosestrife Swamp*—located near the parking lot. In spring, follow the sounds of the marsh and look for Wood Ducks and waders. This is a good place to find Rusty Blackbird and Little Blue Heron. Take your time in this area and follow the path through the woods looking for migrants, woodpeckers, thrushes. If you reach a larger lake near the wetlands, you've arrived at Lake Orbach, which is not part of the preserve. In summer it hosts, on its far side, the privately owned Pouch Boy Scout Camp.

Follow the Blue or Yellow Trail just to the right of Lake Orbach for *Pump House Pond*, a good place for spring and fall migrants. Take advantage of the observation deck for good views.

You can also continue on the Yellow Trail to *Hourglass Pond*, which is not far from Pump House Pond and is fairly marshy. This is a reliable place to find Little Blue Heron.

From this side of the park, doubling back past Lake Orbach, you will also find a path to the fall hawk-watching site *Moses Mountain*, described in Other Places to Find Birds on Staten Island.

The lower portion of the park is easily accessible by trail and includes a freshwater lake. Best advice here is to follow the birds and see where they take you. You shouldn't get terribly lost if you have a map or GPS.

KEY SPECIES BY SEASON

Spring Wood Ducks, Belted Kingfisher, Eastern Phoebe, Blue-headed Vireo, kinglets; over twenty species of warbler, including Black-throated Blue and American Redstart, pass through; Eastern Towhee, Scarlet Tanager, Rusty Blackbird.

Summer Red-eyed Vireo, House Wren, Little Blue Heron, Rusty Blackbird.

Fall Interesting sparrows such as Swamp and Song Sparrows through spring.

Winter Some waterfowl, including Ring-necked Ducks; woodpeckers abound through much of the year.

HABITAT

Dense forest and freshwater wetlands and lakes.

BEST TIME TO GO

Spring through fall are best.

HOW TO GET TO HIGH ROCK PARK AND CONSERVATION CENTER AND MOSES MOUNTAIN

By Bus Take the S57 along Rockland Avenue to the Nevada Avenue stop.
By Car Nevada Avenue dead-ends at the preserve's parking lot.

200 Nevada Avenue, Staten Island, NY 10314
718-667-2165
http://www.nycgovparks.org
http://sigreenbelt.org

There are restrooms near the picnic tables at the Nevada Avenue parking lot. No café.

In winter, some of the trails may be steep and icy.

Knowing bird calls will help at this wild and densely wooded park.

OTHER THAN BIRDING

This is a nature-lover's park, and hiking is the only other activity you'll likely encounter.

Scope Not necessary unless you are visiting Moses Mountain.

Photos Medium when the trees are leafed out.

Mariner's Marsh Park

This park is currently closed to the public while it undergoes chemical remediation. When it reopens, it should return to its status as a real hot spot. With that in mind, we offer this description.

Get lost in this pretty park with lots of ponds and habitats that attract birds. The Lenape Indians camped here and used it as a graveyard. Then it became a bustling industrial area with an iron mill and shipyard. Like many of its fellow parks, it's wedged into a residential development with well-marked trails, which are winding and circuitous. You'll want to have a buddy with you and a working GPS to make your way through the forest and around the ten connected ponds. Each one is a little different and may have different species of birds.

Fall may offer molting Wood Ducks, Belted Kingfishers, and migrating songbirds. Spring will be somewhat productive, but you'll definitely want to pay a visit in early winter before the water freezes for a chance at waterfowl such as Gadwall, Northern Pintail, and Green-winged Teal.

Bridge Creek and Goethals Pond are just a couple of minutes away.

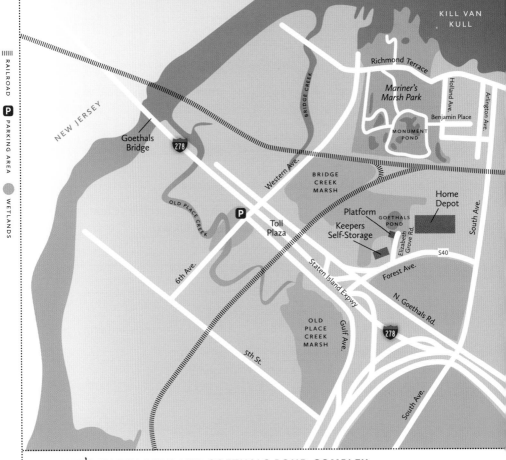

MARINER'S MARSH PARK AND GOETHALS POND COMPLEX

KEY SPECIES BY SEASON

Spring Black-crowned Night-Heron, Least Bittern, Brown Thrasher; migrants including Common Yellowthroat, Yellow, Blackburnian, and Black-throated Blue Warblers; Orchard and Baltimore Orioles.

Summer Yellowlegs, Belted Kingfisher, breeding woodpeckers, songbirds and more common birds.

Fall Late fall at Mariner's has Wood Duck and other arriving waterfowl, including Gadwall; Belted Kingfisher, woodpeckers and sparrows.

Winter Overwintering waterfowl including Northern Pintail, Gadwall, Green-winged Teal; backyard birds; Red-tailed and Sharp-shinned Hawks.

167

Freshwater ponds and wetlands, old-growth and transitional woodland.

BEST TIME TO GO

Fall and early winter for waterfowl, woodpeckers, and sparrows.

HOW TO GET TO MARINER'S MARSH PARK

Located in the northwest section of Staten Island; entrance information may change when park reopens, so call or check ahead.

INFORMATION

Richmond Terrace, Staten Island, NY 10303
212-639-9675
http://www.nycgovparks.org

No restrooms or café.
Check for when this park opens to the public.
   Scope Wouldn't help.
   Photos Easy.

Goethals Pond Complex, Including Bridge Creek, Old Place Creek Park, and Goethals Pond

In northwest Staten Island, this diverse wetlands area near the Goethals Bridge toll is worth a couple of quick stops. The complex is a series of fresh and tidal marshes and shallow freshwater that has Goethals Pond as its focus. The other nearby spots might have some interesting birds, and since you will probably have to drive past them anyway, why not take a moment and satisfy your curiosity? The marshes attract migrating shorebirds and offer critical breeding habitat for a number of other birds, although most of these places are not pristine.

VIEWING SPOTS

   Bridge Creek Along Western Avenue between Goethals Road North and Arlington Yards, you'll find easy right-from-the-road viewing across twenty-two acres of restored marshland with a gas pipeline running through it. Luckily the wading birds and shorebirds seem not to mind. Wilson's Snipe have been seen here as well. Pull over to the side of the road and park at any location that seems to have good views. In spring and summer find Great and Snowy Egret, yellowlegs, and Killdeer, as well as Belted Kingfisher, Song Sparrow, and a possible warbler or two. In winter you may see Hooded Merganser and American Black Duck.

Old Place Creek Park To visit this marsh, stay on Western Avenue heading south, and just after you go under Highway 278 make a stop and look around at Sixth Avenue to view seventy critical acres of tidal marsh. This is a fairly industrial area, but wading birds can be found working the pond. There is a small parking lot, and if you want further access, a number to call is posted on the chain-link fence. Look for Snowy and Great Egrets, yellowlegs, Song Sparrow, and Red-winged Blackbird, as well as overwintering Green-winged Teal, American Black Duck, Gadwall, and Northern Pintail.

To reach Goethals Pond itself, make a left onto Forest Avenue and then a left onto Elizabeth Grove Road, a narrow street just past the mini-storage. (If you get as far as the Home Depot, you have gone too far.) Park in the lot and start scanning the bushes. Register your visit, and follow the path to the elevated platform. The pond can have up to three feet of water, or it can be a mudflat. It's a good but trashy place to bird in spring through late summer and fall, and can be a hot place for herons and shorebirds like Greater and Lesser Yellowlegs; Solitary, Semipalmated, and Least Sandpipers; Pectoral and the occasional Stilt Sandpipers, Short and Long-billed Dowitchers, Common Gallinule, and in spring, the chance of a glimpse at Wilson's Snipe. In summer, Kildeer and Spotted Sandpiper breed here. Summer also offers views of Little Blue and Green Heron, Marsh Wren, Yellow Warbler, Common Yellowthroat, Eastern Towhee, Red-winged Blackbird, Common Grackle, Barn Swallow, and American Goldfinch.

BEST TIME TO GO

Spring and fall migration are good; summer is good for nesting birds and some shorebird migration.

HOW TO GET TO THE GOETHALS POND COMPLEX

If you want to see the entire complex, a car is necessary. The S40 bus stops near the Goethals Pond viewing area.

Goethals Pond is located north off Forest Avenue between the Home Depot and Keepers Self Storage on Elizabeth Cove Road.

Bridge Creek is located off Western Avenue between Goethals Road North and Arlington Yards.

INFORMATION

Old Place Creek Park and Goethals Pond Complex are described at http://www.dec.ny.gov/outdoor

No restroom facilities and no cafés. However, there are places to eat in the area.

Be alert, as Goethals Pond can be a lonely place. Best to go with a pal.

Scope Yes, at Goethals Pond.

Photos Easy.

Snug Harbor and Allison Pond Park

One of the loveliest spots to visit on Staten Island, Snug Harbor was originally designed in the nineteenth century as a retirement settlement for merchant seamen. Gorgeous Greek-revival buildings face the Kill van Kull. A freshwater lake, botanical garden, and an authentic Chinese Scholar's Garden are all part of this unusual complex. This might not be the top spot to see birds on Staten Island, but a trip here is soul restoring, and it's easy to reach from the Staten Island Ferry, making for a memorable outing for birders and non-birders alike. Allison Pond Park is also an interesting spot near Snug Harbor, and together they make for a good combo.

VIEWING SPOTS

Your first stop is Snug Harbor. Check the lawns and trees in front of the buildings facing the water for migrants and sparrows. Follow the road from the parking lot back to the botanical garden, which is good for birds and possibly even better for butterflies. Waterfowl and some waders may be found on the freshwater pond and Barn Swallows in summer. And don't miss the Chinese Scholar's Garden nearby and its authentically re-created buildings and gardens. There is the occasional heron fishing for carp in the ponds, and the garden itself is a wonderful surprise. This beautiful location is not heavily visited and can be a peaceful spot to find birds.

Snug Harbor doesn't take all day to see, so you might consider making

White-throated Sparrow

nearby *Allison Pond Park* your next stop. Leave Snug Harbor via central Cottage Row, cross Henderson Avenue, and walk one block to Prospect Avenue to find this picturesque little park.

In spring, expect to see migrants, herons, waders, and hawks. The best spot to bird is the stream south from Allison Pond. Also walk around the pond and visit the wooded area and grassy fields for a complete picture.

HOW TO GET TO SNUG HARBOR

Snug Harbor is a two-mile walk or short cab ride from the Staten Island Ferry.

By Bus Take the S40 and stop at Richmond Terrace / Sailors Snug Harbor Gate.

Willowbrook Park

This Greenbelt park is one of Staten Island's most popular and remains a good location for birds as well, with its dense forests and lake. Over one hundred species of birds have been found here. Locals know it as a migrant trap where you can see fifteen to twenty species of warbler in one day during migration, and consistent levels of activity spring through fall. If you are looking for a Greenbelt location that is a bit off the normal rounds for birders, try this one.

Enter from Eaton Place and check out Willowbrook Lake for waterfowl like Green-winged Teal fall through spring, plus Belted Kingfisher, sandpipers, and some waders. Take the White Trail for the swamp forest area at the park's farthest reaches. This may be the best spot to visit in summer, as you can expect all local Staten Island nesters, including Warbling Vireos, on the edges, plus Eastern Towhees, Yellow Warbler, Baltimore Oriole. It's also the best place on Staten Island to see Rusty Blackbirds. Check the swamp woodland for wrens and the occasional Summer Tanager.

HOW TO GET TO WILLOWBROOK PARK

By Bus Take the S62, S92, X10, X10B, or X11 to the Victory Boulevard/Morani Street stop. Or take one of the many buses that service the Eton Place stop: S44, S59, S94, X17, X17A, X17J.

By Car There are parking lots accessed via Eton Street or Victory Boulevard.

INFORMATION

1 Eaton Place, Staten Island, NY 10314
718-698-2186
Restrooms are by the parking lot.

Miller Field, Midland Beach, and South Beach

When combined in one visit, the beaches and open ground at these locations can provide a nice diversity of birds—and maybe some unusual ones as well. Make your priority Miller Field, which originally was a freshwater wetland, later a farm, and then an Army Air Corps base in the early years of aviation. Now its 187 acres provide athletic fields and a habitat for grassland birds. Midland Beach and South Beach offer some gulls and a view out onto Raritan Bay for winter waterfowl.

VIEWING SPOTS

Miller Field is good for sparrows and other grassland specialties, which might include Horned Lark or Snow Bunting in winter. Lapland Longspur have been seen.

In front of Miller Field is a beach, and if you make a left and walk north, you will come to Midland and then, despite walking northeast, South Beach. Use the two-and-a-half-mile-long boardwalk that runs parallel between the water and Father Capodanno Boulevard. Plenty of parking lots service this popular summer Coney Island–like area, but you'll be interested in going in the winter to view waterfowl like Long-tailed Duck and Red-breasted Merganser. Occasionally there are more interesting sightings like Snow Geese and White-fronted Goose. Purple Sandpipers may also visit, along with a variety of other shorebirds. During a winter irruption, the black pines may have crossbills.

HOW TO GET TO MILLER FIELD

By Bus Take the S76 to any of the stops on New Dorp Lane.

By Car Park at the end of New Dorp Lane to enter Miller Field and the beaches.

INFORMATION

The paths can sometimes get muddy.

Restrooms are open year-round at the Fountain of the Dolphins / Sand Lane, at the fishing pier near Seaview Avenue, and at the south end of the beach near Greeley Avenue.

OTHER THAN BIRDING

There are lots of sports facilities—ball fields, archery range, tennis courts—and the Carousel for all Children.

OTHER PLACES TO FIND BIRDS ON STATEN ISLAND

Staten Island is full of wild areas and small, attractive parks. None of the following is a primary hot spot, but each has its own character, and they are all small enough that you can do them as short jaunts individually or several in a day. The most productive are marked with an asterisk.

King Fisher Park

These twenty-three acres of woodlands also include a freshwater pond. The woods are a warbler haven spring through fall, while the pond may turn up some interesting waterfowl in winter and migrating waterbirds otherwise. Parking and the entrance gate are on Fairfield Street across from PS 37.

Oakwood Beach

Next to Great Kills Park, this beach and tidal marsh complex is where local birders make a quick stop for a wide range of birds, especially from spring through fall. Walk down from Mill Road and enter at the south end of Tarleton Street next to the Oakwood Beach Water Pollution Control Plant. In summer, check at low tide for herons, oystercatchers, and skimmers working the shoreline. Go to the concrete bulkhead and scan over the bay. The tidal marshes have Red-winged Blackbirds, Boat-tailed Grackles, Killdeer, and waders in spring. It's also a decent spot to find winter birds, including gulls. Popular during the Christmas bird count.

Moses Mountain

This two-hundred-foot hill, created with the debris from the excavation of the never-completed Richmond Parkway, is the best spot on Staten Island to watch hawks in the fall. Cooper's, Sharpies, and the occasional Bald Eagle sail over this Greenbelt site. Golden Eagles have been seen. You can also watch songbirds, like tanagers and finches, on migration. Reached via Manor Road across from High Rock Park, or by a trail through High Rock.

Fort Wadsworth*

This historical military complex is located at an outcropping on Staten Island's northeast point. Bring your scope, and this could be a convenient and quick stop in the winter for easy views of waterfowl. Two parking lots, with a restroom facility in between (which may be closed in the off-season) get you pretty close to your destination, a cove just north of the point.

You can usually find the most common saltwater ducks like Bufflehead and Greater Scaup, but a visit here may also yield Horned Grebe, Red-necked Grebe, American Wigeon, Common Goldeneye, and Great Cormorant. In fall, it's a nice place to look for sparrows. While here, check out the jetty for Purple Sandpipers, while the flats may also produce a pleasant surprise. Don't forget to scan the Verrazano Bridge, as it is home to a pair of Peregrines. If you have more time, take a walk on *South Beach* for more waterfowl in the ocean and Lesser Black-backed Gulls.

HOW TO GET TO FORT WADSWORTH

By Bus Use the S51 or S81, both of which can be picked up at the ferry terminal in St. George.

By Car Set your GPS for the corner of Battery Road and USS North Carolina Road for the bigger of the two parking lots. To get down to the beach, pick up Range Road off USS North Carolina Road.

Silver Lake Park

This large freshwater lake surrounded by a golf course and playgrounds is actually a reservoir and doesn't have the woods and other attractive features you would hope to find in a natural area. This is not a highly recommended stop, but if you're at adjacent Clove Lakes or in the Greenbelt, have your scope handy, and want to be thorough, why not?

The best time to visit is winter for waterfowl. Spring and fall have some songbirds, and in fall you can look for migrating raptors at the bridge. Ruddy Ducks in particular like this spot, and there can be nice-size rafts of them. You can also see other waterfowl, including Hooded Merganser and Pied-billed Grebe. Take a quick spin by in summer and watch Barn and Tree Swallows working the insects over the reservoir.

Limited parking is available on Victory Boulevard and Forrest Avenue. The Silver Lake Golf Course at 915 Victory Boulevard is open to the public and has a restaurant.

http://www.nycgovparks.org

Tottenville Train Station

We are told that Eurasian Wigeon come here every year, which alone seems like a good reason for a stop in winter. Also check this saltwater harbor in winter for scoters, Gadwall, mergansers, and Horned Grebes. If you take the train from the Staten Island Ferry to its terminus at Tottenville, any extra stops you make along the way are at no charge. You can drive there as well—it's at the southern end of Staten Island. Take a scope.

The Greenbelt hums with activity spring through fall—waders, thrushes, Baltimore Oriole, Eastern Towhee, and many others find the woodlands and swamps good for migrating through or nesting.

Reed's Basket Willow Swamp*

Located slightly beyond the northeast contiguous parkland of the Greenbelt, this forested area is part of a protected wetland with several freshwater ponds. It might be interesting during migration for songbirds and in summer for waders. It can be reached by car via Spring Street as it turns into Forest Road's dead end.

LaTourette Park

Sharing its space with a golf course, LaTourette is one of the largest Greenbelt parks. Woodsy Buck's Hollow in the northern section might be an interesting spot for woodpeckers, migrants, and sparrows. There is a Nature Center and parking off Rockland Avenue and a bus stop for the S54 for Rockland Avenue/Manor Road. Blue, Red, Yellow, and White Trails will take you through most of LaTourette. Head south on the Blue and Multi-Purpose Trails to Old Mill Road and the Field of Dreams near St. Andrews Church, where you can find Indigo Buntings and where Orchard and Baltimore Orioles nest.

For a quick look at a freshwater environment at the west end of the Greenbelt, stop on Richmond Avenue and survey Richmond Creek, accessed via Richmond Avenue between Arthur Kill and Forest Hill Roads.

IN CASE YOU WERE WONDERING

Freshkills Park

Reclaimed landfills often make for great birding, and Freshkills is no exception. Formerly the world's largest landfill at twenty-two hundred acres, Freshkills is closed to the public most of the time but is slowly undergoing a multiphase conversion to a park that will provide wild areas, waterways, and public grounds. Until that ambitious plan is fully realized, you can try to visit by applying for permission from the Parks Department or just check for birding tours offered by the Freshkills Park Alliance, http://freshkillspark.org.

Harbor Herons Complex—Shooters and Prall's Islands

These two uninhabited islands off the north and west coasts, respectively, of Staten Island are unreachable by anyone except researchers, and they are part of an Important Bird Area that protects colonial nesting waders and waterbirds. A series of disastrous events from an oil spill, a hurricane, and the intrusion of invasive plant and animal species have for now left Prall's without nesting birds. An ongoing cleanup and restoration will hopefully bring the islands back to providing the much-needed protected habitat for these birds of concern.

Staten Island Ferry

The iconic Staten Island Ferry is a major tourist attraction, and you would think it would be a terrific way to see birds in the harbor, but sadly, it is not. It does provide spectacular views of the Financial District, Governors Island, and the Statue of Liberty, and is a wonderful way to spend thirty minutes. During the ride, you may get some cormorants and a few gulls behind the boat, but you are often required to remain inside, making any viewing difficult. If you take this free ferry from Manhattan to bird Staten Island, the only hot spot close to the terminal is *Snug Harbor*—a short taxi ride or maybe a twenty-minute walk away.

6   Nassau County

Once part of Queens County, Nassau became independent in the late nineteenth century when Queens joined New York City. Now it occupies a strip of suburban Long Island that stretches from the Sound to the Atlantic Ocean and is bounded on the east by Suffolk County.

In the early 1900s, the North Shore was a haven for wealthy New Yorkers, earning the nickname "the Gold Coast." Theodore Roosevelt had a summer home here at *Sagamore Hill*, and like other former great estates such as those at *Sands Point Preserve* and the *Bailey Arboretum*, it is now open to the public. These preserves are satisfying stops for both birders and non-birders. In this category, *Muttontown Preserve* is an often overlooked gem, with leafy hiking trails and abundant birds. The *North Shore* also has well-preserved wetlands such as *Shu Swamp*.

But the *South Shore* in general has better birding sites and includes several Important Bird Areas and major hot spots. *Jones Beach State Park* should be on your not-to-miss list in Nassau County. It has a wild beach area away from the crowds where birds can nest and gain refuge year-round. Nearby are *Point Lookout*, a winter waterfowl haven, and *Nickerson Beach*, which hosts lots of easily photographed beach-nesting birds in summer. *Cow Meadow Park* and *Oceanside Marine Study Nature Center* provide wetland environments. *Hempstead State Park* is a place to go for spring migration, and the lakes and woodlands at pretty *Massapequa Preserve* can be enjoyed any time of year. The South Shore also has a series of freshwater ponds that were originally created for drainage but have become winter waterfowl havens. They offer close-up views of freshwater ducks, and we highlight a couple of spots that are off the radar on this tour with *Camman's Pond, Grant Park Pond*, and tiny *Willow Pond*.

You will need a car to visit nearly all of these sites, but that's life in Nassau County. A few are in the Long Island Greenbelt's more than two hundred miles of trails, allowing you to hike in to your destination.

LONG ISLAND SOUND

95

Caumsett State Historic Park

Centre Island Town Park

Ransom Beach

Stehli Beach

BAYVILLE

Sagamore Hill

Welwyn Preserve

Bailey Arboretum

Shu Swamp

OYSTER BAY

HUNTINGTON

Sands Point Preserve

GLEN COVE

Upper Francis Pond

Lower Francis Pond

Mill Pond

OYSTER BAY

Garvies Point Preserve

101

Planting Fields Arboretum

25A

Uplands Farm Sanctuary

St. John's Pond Preserve

PORT WASHINGTON

Leeds Pond Preserve

Muttontown Preserve

SYOSSET

25

William Cullen Bryant Preserve

106

GREAT NECK

107

Whitney Pond Park

135

Northern Blvd.

Long Island Expwy.

295

495

110

SUFFOLK

25B

MINEOLA

25

GARDEN CITY

Meadowbrook State Pkwy.

Wantagh State Pkwy.

LEVITTOWN

QUEENS

25

Southern State Pkwy.

Tackapausha Preserve

Sunrise Hwy.

Hempstead Lake State Park

Twin Lakes Preserve

Massapequa Preserve

27

Sunrise Hwy.

BALDWIN

27

Mill Pond Park

FREEPORT

JFK Airport

Grant Pond Park

Lofts Pond Park

Milburn Pond Park

Willow Pond

HEWLETT

OCEANSIDE

Cow Meadow Park and Preserve

Camman's Pond Park

JAMAICA BAY

878

Oceanside Marine Nature Study Area

Tobay Beach

Gilgo Beach

Ocean Pkwy.

LONG BEACH

Loop Pkwy.

Point Lookout

Nickerson Beach

Jones Beach State Park

John F. Kennedy Memorial Wildlife Sanctuary

NASSAU COUNTY

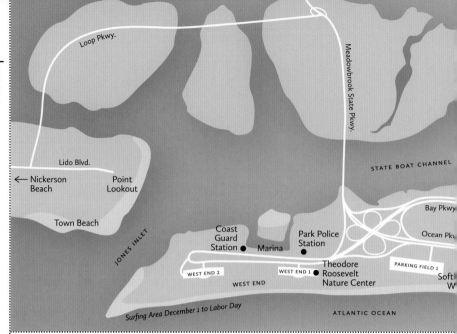

Loop Pkwy.

Meadowbrook State Pkwy.

Lido Blvd.

STATE BOAT CHANNEL

← Nickerson
Beach

Point
Lookout

Bay Pkwy

Town Beach

JONES INLET

Coast
Guard
Station ● Marina

Park Police
Station
●

Ocean Pkw

WEST END 2

WEST END 1 ●

Theodore
Roosevelt
Nature Center

PARKING FIELD 1

Soft
W

WEST END

Surfing Area December 1 to Labor Day

ATLANTIC OCEAN

JONES BEACH STATE PARK

KEY SITES

Jones Beach State Park

Often considered to have the best birding in all of Nassau County, with a large number of species and an impressive list of rarities, Jones Beach State Park can be counted on as a terrific place to find birds in any season. At six and a half miles long, this Important Bird Area is massive and so diverse that it warrants a full-day visit on its own. Jones Beach is a man-made landscape created by dredging sand to connect several smaller islands, and park commissioner Robert Moses designed the park so that it had a beach large enough to accommodate enormous crowds. Indeed, it is a hugely popular summer beach destination, with six to eight million visitors each year on the sand, in the surf, and on the two-mile-long boardwalk. If you go on a summer weekend, it's a given you'll be jostling among thousands of beachgoers.

VIEWING SPOTS

The scene at Jones Beach is dominated by the Water Tower, a two-hundred-foot-high brick-and-stone campanile-style obelisk. Use it as a marker for orienting yourself around this immense park. For the most rewarding expe-

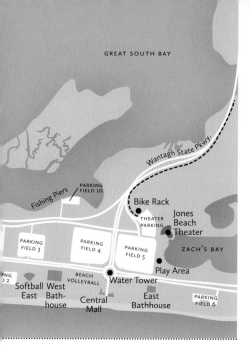

Dunlin

rience any time of year, you can't beat the West End for its wild beaches and natural sand dunes. Located past the end of the boardwalk, it also includes the Coast Guard station and the park's westernmost tip.

In winter, you may have the place more or less to yourself, so enjoy the quiet and check out *Zach's Bay*, by the Nikon Jones Beach Theater, for waterfowl. The fields and numbered parking areas sometimes harbor Horned Larks and Snow Buntings. Also visit the *West End*, the area surrounding the *Coast Guard Station and marina*. Stands of trees will have overwintering songbirds like Yellow-rumped Warbler and American Tree Sparrow. If you're lucky, rare finches may be present, like Red or White-winged Crossbills, Common Redpolls, or Pine Siskins. From the West End 2 parking lot, hike out over the sand to the *westernmost tip* for more waterfowl and better chances of seeing Northern Gannets (a year-round presence) on the ocean side. These *dunes* are also known for their Merlin and Snowy Owls. There may be large concentrations of birds in the waters, and they come relatively close to shore. You can also scan out across Jones Inlet to Point Lookout and north toward the Loop Parkway. Look for *open sandbars at low tide* where seals may haul out in winter; and in summer, they are where terns, plovers, Red Knots, and other shorebirds rest. A thorough trip around the park on a good winter day will net over a dozen water species and could include scaup, scoters, loons, and a nice assortment of bay ducks.

Spring brings migrating shorebirds that feed and lounge on the beach and on the *bayside sandbars* and are best seen at low tide. Later in the season, you'll find songbirds that like the trees, brush, and mowed areas. The *median* above the West End parking lot is another good spot to scan. Park your car in one of the West End lots and walk out to the Bay Parkway to see what's going on inside that loop.

Although this is not known as a spot to view nesting shorebirds, if you do make the trip in summer to see the few Piping Plovers, American Oyster-catchers, Least Terns, and Black Skimmers that nest on the West End beaches, hike out to the tip. Go before the crowds arrive in the morning or after they leave for the day. From April through most of August, the dunes are roped off to provide a safer nesting environment. Please obey the rules and give the parents a wide space in which to raise their chicks. In late summer, check these beaches for unusual terns like migrating Black Terns.

After the skimmers and plovers leave, there will still be oystercatchers. Fall is also a great time to be on the beach looking for migrating shorebirds and hawks. Songbirds return as well and bring with them a greater assortment of sparrows. If you come in September or October you may find yourself swept up in a swirl of Tree Swallows.

KEY SPECIES BY SEASON

Spring Shorebirds migrate through, although fall is better. You'll have a chance at Gull-billed Tern. Migrating songbirds drop down, along with raptors like Northern Harrier, Peregrine Falcon, and, uncommonly, Rough-legged Hawk. Short-eared Owl have been seen. Eastern Towhee arrive and leave in the fall.

Summer Even though the park can be packed with people, come early or late in the day to appreciate Red Knot, Piping Plover, American Oyster-catcher, Least Tern, Black Skimmer, Osprey, Great Egret, Glossy Ibis, Willow Flycatcher, Tree and Barn Swallows, Brown Thrasher, Common Yellowthroat, Yellow Warbler, and Boat-tailed Grackle.

Fall A terrific spot for migrating shorebirds, including some uncommon-to-rare ones like the White-rumped Sandpiper, Red Knot, and American Golden-Plover. Migrating songbirds are slightly better in fall than spring. Fall is the best time for sparrows, including the chance to see unusual ones

Green-winged Teal

like White-crowned, Clay-colored, and Vesper. Fall also brings Royal Tern and Northern Harrier, plus migrating Tree Swallows. You could find a few additional raptors, but the hawk watch at Fort Tilden is recommended over Jones Beach.

Winter Superior location for waterfowl, including Surf, White-winged, and Black Scoters; Common Goldeneye, Harlequin and Long-tailed Ducks, Common Eider, Common and Red-throated Loons, Horned Grebe, Northern Gannet, Razorbill. Overwintering shorebirds like American Oystercatcher, Sanderling, Black-bellied Plover, Dunlin, Purple Sandpiper; Merlin and Peregrine Falcon; uncommonly, Bonaparte's Gull; Snowy Owl during irruptions; Short-eared, Long-eared, and Saw-whet Owls make occasional appearances. Songbirds include Yellow-rumped Warblers, Horned Larks, and Snow Buntings, and rarely, American Pipit and Lapland Longspur. Some sparrows remain and are sometimes joined by Pine Siskin and Common Redpoll; Red and White-winged Crossbills in irruption years.

HABITAT

Beach, marine, grassland, and scrub.

BEST TIME TO GO

A year-round preserve, but winter, spring, fall, summer, in that order.

NASSAU COUNTY

HOW TO GET TO JONES BEACH STATE PARK

From the west, take the Meadowbrook State Parkway to Ocean Parkway, or from the east, take the Wantagh State Parkway, which becomes the Jones Beach Causeway.

There are several parking lots, but you can park all day at the Coast Guard Station and have access to the places that you most want to visit.

INFORMATION

1 Ocean Parkway, Wantagh, NY 11793
516-785-1600

Food is not available in the off-season; not all restrooms are open year-round.

There may be a fee to park, but if you go in the early morning or late afternoon, or in the off-season, you may not be charged.

The Theodore Roosevelt Nature Center has family and adult programs.

OTHER THAN BIRDING

Jones Beach has a full array of beach activities in the summer, including swimming, surfing, fishing, concerts, softball, volleyball, and many other activities. Seal walks take place in winter.

Scope Necessary.

Photos Bring a long lens.

Harlequin Duck

Point Lookout

One of a pair of exceptional birding locations west of Jones Beach, Point Lookout is a reliable winter viewing area for waterfowl on the west side of Jones Inlet and an Important Bird Area. Visit here for the oceanfront beach and jetties, which attract large numbers of gulls and waterfowl from November through early April. Don't forget the scope!

VIEWING SPOTS

From your street parking spot at the end of Lido Boulevard, walk east to the inlet. If you walk to the right down the beach, you will find the jetties. Keep walking south until you round the point and continue west until you reach the Town Park. It is a private beach, but during the winter no one will make you feel like you are trespassing. This is a good place to view Northern Gannets and grebes.

KEY SPECIES BY SEASON

Winter The jetties provide a haven for waterfowl like Harlequin Duck, scaup, rafts of Brant, Northern Pintail. You'll also find Northern Gannet and Common Eider in decent numbers, sometimes Long-tailed Duck in the hundreds; Common and Red-throated Loons and a variety of other waterfowl can all be seen bobbing about. The beach is known to host a wide variety of gulls in the winter, and Iceland Gulls are occasionally seen here.

Spring There may be Harlequin Duck, loons, Horned and Red-necked Grebes, and other waterfowl through early April; Bonaparte's Gull.

Fall Some warbler action; sparrows; flocks of Black-bellied Plover and other migrating shorebirds.

HABITAT

Beach and marine.

BEST TIME TO GO

November through March.

HOW TO GET TO POINT LOOKOUT

Take the Meadowbrook Parkway to the Loop Parkway and have your birding buddy keep an eye out for birds in the marsh. Turn left on Lido Boulevard at

Least Terns mating

the light and travel straight on through the town of Point Lookout. There is street parking only here, so look for a legal spot that is not reserved for residents. Walk to the end of Lido and enter the beach there. The inlet will be on your left, jetties on your right.

INFORMATION

Town Park at Point Lookout: 516-431-3900

In winter the beaches are free, but between Memorial Day and Labor Day there are fees.

No public restrooms. Restaurants in the area are open year-round.

Scope Needed.

Photos Birds may be at a distance, so a long lens is necessary.

Nickerson Beach

The second location close to Jones Beach is Nickerson Beach, a particularly good place to visit in the summer for nesting Piping Plover, Least and Common Terns, American Oystercatcher, and an impressive Black Skimmer colony. The birds can be easily seen and photographed, and being in the presence of large nesting colonies like the skimmers can be an extraordinary experience.

VIEWING SPOTS

Located just west of Point Lookout, Nickerson is a popular beach destination, and from May to September you can watch birds raise their chicks on the beach in front of the parking lot and further west. Look for the roped-off areas near the dunes protecting the nests.

Because it is so heavily visited in the summer, the birds have to work around the crowds, adding another level of stress to their already complicated lives. Be considerate of their needs and aware that the off-hours, when birders are likely to visit, are the only remotely safe time the birds have on the beach with their chicks. Please be respectful when you are viewing and taking photographs.

KEY SPECIES BY SEASON

Summer Beach nesters include Least and Common Terns, Piping Plover, and American Oystercatcher; Killdeer, Willet, and other shorebirds visit. You may also see Roseate and Forster's Terns, and the odd rarity. Black Skimmers nest later in the summer.

Fall Shorebirds migrate through, but there's not as much to see as there is during breeding season.

HABITAT

Beach and marine.

BEST TIME TO GO

May to September.

INFORMATION

880 Lido Boulevard, Lido Beach, NY 11561
516-571-7700

Campground, playground, beach cabanas, and food concession stand are all open only during the summer. There are hefty parking fees Memorial Day through Labor Day. You can pay the fee, buy the pass, or try your luck by going very early in the morning or late in the day after normal beach hours.

Scope Not necessary.

Photos Great place to photograph nesting shorebirds.

Cow Meadow Park and Preserve

Don't pass up the opportunity to bird this multiuse park with ball fields and courts, playground and marina on the South Shore in Freeport. It's easy to access and quite appealing—both visually and for birds. Cow Meadow Preserve comprises 150 acres of Long Island's marine wetlands, including salt marsh, mudflat, and tidal creek habitats. More than 150 species of birds can be found here, from waders and waterfowl to woodpeckers and migrants.

VIEWING SPOTS

Leave the parking area and walk toward the pond visible from the lot. In winter this is a good place to see waterfowl as well as raptors, so don't forget to check out the trees around the pond. Follow the path along the right of the pond, and with a little effort you can make your way the short distance to the marsh and beach. Once at the beach, walk along the pretty grassy area and find an Osprey platform as well as a Bluebird Trail. Waders and shorebirds are found in the clear water along the shoreline and on the sandbars.

KEY SPECIES BY SEASON

Spring Killdeer, Song and Saltmarsh Sparrows, Yellow Warbler, Red-winged Blackbird.

Summer Great Blue Heron, Great and Snowy Egrets, Yellow and Black-crowned Night-Herons, Glossy Ibis, Osprey, Black-bellied Plover; Solitary, Least, and Semipalmated Sandpipers; Willet, Short-billed Dowitcher, Laughing Gull, Common and Forster's Terns, Black Skimmer, Cedar Waxwing, Yellow Warbler, American Goldfinch; Clapper Rail have been seen.

Fall Arriving waterfowl including Green-winged Teal, Hooded Merganser; Greater and Lesser Yellowlegs, Great Blue Heron, Yellow-rumped Warbler, White-throated Sparrow.

Winter Waterfowl including Green-winged Teal and, much less commonly, Blue-winged Teal; scaup, Brant, Ruddy Duck, Hooded Merganser, Double-crested Cormorant, Downy Woodpecker, Northern Flicker, and Black-capped Chickadee.

HABITAT

Salt marsh, mudflats, tidal creek, beach, and forest edge.

There is something of interest year-round, but the most popular times with birders are summer, late fall, winter, and early spring.

HOW TO GET TO COW MEADOW PRESERVE

Take the Southern State Parkway to Exit 22 South (Meadowbrook Parkway). Exit at M9 West (Merrick Road). Proceed to the left lane immediately and turn left onto Mill Road. Take Mill Road to South Main Street and turn left. Proceed down South Main Street to the park. The entrance is on the left.

If you are making a tour of Nassau County's winter waterfowl stops and coming from Milburn Pond in Freeport, take West Merrick Road east and make a right on South Main Street. The entrance is on the left.

INFORMATION

South Main Street, Freeport, near Ann Drive South
516-571-8685
https://www.nassaucountyny.gov

Much of the preserve is off-limits, but there are plenty of things to see from the quarter mile of trails, the parking lot, and on the beach. However, this is a preserve that you should not visit alone.

OTHER THAN BIRDING

Athletic fields, chess and checker tables, handball, picnic areas, a playground, and a fishing pier.

Scope Not really necessary.

Photos Easy.

Oceanside Marine Nature Study Area

In the 1970s these fifty-two acres of pristine wetlands were set aside from development and designated a Marine Nature Study Area. The preserve offers many ways to learn about wetlands, as well as excellent birding opportunities in this carefully stewarded, rich marine marsh. Although this is a relatively small preserve, over two hundred species have been recorded here, and the sanctuary is easily explored using the one mile of trails and boardwalks.

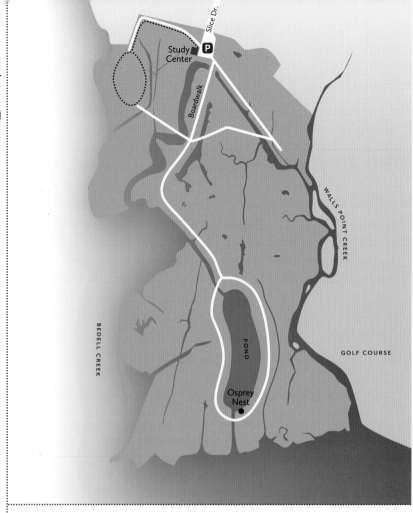

OCEANSIDE MARINE NATURE STUDY AREA

Highly recommended, especially in summer when birdlife is at its fullest and when you have an excellent chance of seeing salt marsh nesters like Marsh Wren and Saltmarsh and Seaside Sparrows.

VIEWING SPOTS

For the best views, plan to visit at low tide. At the parking lot, stop by the Study Center and look at reports of recent species. Take the boardwalk across the marsh in any direction; the one that heads out straight to the Osprey nest and loops around may be the most satisfying, but you really can't make a wrong choice.

KEY SPECIES BY SEASON

Spring Osprey arrive and stay to nest; migrating shorebirds; some songbirds.

Summer An abundance of wading birds, including Great Egret and a chance to see Little Blue and Tricolored Heron, Glossy Ibis; Clapper Rail, Sora; Greater and Lesser Yellowlegs, Willet; Laughing Gull, Common and Forster's Terns; Black Skimmer, American Oystercatcher; Willow Flycatcher; Cedar Waxwing; Saltmarsh Sparrow (year-round) and Seaside Sparrow; Marsh Wren; Baltimore Oriole.

Fall Migrating shorebirds begin to arrive in July and can include the more unusual ones like Stilt, White-rumped, and Western Sandpipers; some songbirds.

Winter Waterfowl such as Brant, Northern Pintail, American Wigeon; possibly Bonaparte's Gull; Dunlin are present in spring and fall as well; Fox and American Tree Sparrows all winter.

HABITAT

Tidal wetlands.

BEST TIME TO GO

Recommended year-round, but prettiest and best in summer. Accessible only by car.

Willets

500 Slice Drive, Oceanside, NY 11572
516-766-1580
http://mnsa.info/fauna_flora/birdchecklist.htm provides a bird checklist
http://www.townofhempstead.org

Closed on Mondays, when you cannot get into the parking lot or gain access
to the trails.

The Marine Nature Study Area is ideal for a family outing. Not only are
there opportunities to see the salt marsh up close and personal, but the
birds are relatively easy to find. It's great for both birders and non-birding
naturalists.

Restrooms in the Study Center; picnic tables available.

Scope Useful in winter.

Photos Easy.

Hempstead Lake State Park

Several lakes and ponds surround Hempstead Lake, the largest in Nassau
County. The focus here has been fishing, but over the years a number of sports
and other outdoor activities have been added to the park. These changes
have likely had an impact on the numbers and diversity of the species found
in this designated Important Bird Area. The birding is best in the spring
during warbler migration, and Hempstead offers the most consistent and
best spring songbird watching in Nassau County. It's also when regulars
comb the park for rarities.

VIEWING SPOTS

Bypass the main parking lot as you enter from the north on Lakeside Drive,
as well as the second parking lot that services the tennis courts, and park
in the lot at the southern end of the park. Explore the trails around here,
and do not forget the three little ponds nearby—the one on your right as
you're leaving the lot heading south, MacDonald's Pond, and South Pond.
If you still have time, double back and cross Lakeside Drive to bird around
Hempstead Lake itself.

A visit to *South Pond* on or about April 9 often nets a variety of swallows.
You may be tempted to park your car on the streets in the residential neigh-

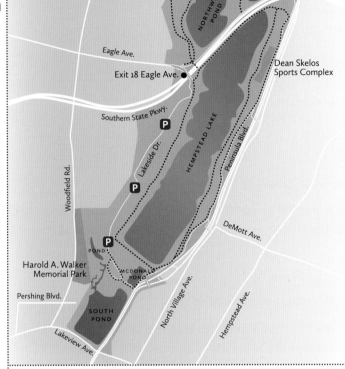

NORTHWEST POND

NORTHEAST POND

Eagle Ave.

Exit 18 Eagle Ave.

Dean Skelos
Sports Complex

Southern State Pkwy.

P

Lakeside Dr.

HEMPSTEAD LAKE

Peninsula Blvd.

P

Woodfield Rd.

DeMott Ave.

P

POND

Harold A. Walker
Memorial Park

McDONALD
POND

Pershing Blvd.

North Village Ave.

Hempstead Ave.

SOUTH
POND

Lakeview Ave.

HEMPSTEAD LAKE STATE PARK

borhood west of the pond, along Pershing Boulevard, and walk in from there. In the morning, viewing the birds will be frustrating from this direction as the sun is coming from the east, backlighting your quarry. Better to leave your car in the southern parking lot within the park as described above and hike in on the Hike Bike Trail, where you'll be able to view South Pond from the north and east. Note that it is not possible to pull over and stop along Peninsula Boulevard, the busy thoroughfare bordering the park's eastern side.

KEY SPECIES BY SEASON

Spring All five migrating swallows have been recorded here: Northern Rough-winged, Tree, Bank, Barn, and, with less regularity, Cliff. In the first weeks of May, come for the warblers and other songbirds.

Summer Nesters include Great Crested Flycatcher, Blue-gray Gnatcatcher, Yellow Warbler, American Redstart, Chipping Sparrow.

Fall Returning songbirds.

Winter Some waterfowl and overwintering birds like chickadee and woodpeckers.

HABITAT

Forest, marshland, and freshwater ponds.

BEST TIME TO GO

Spring.

HOW TO GET TO HEMPSTEAD LAKE STATE PARK

There is a quick-access entrance at Exit 18 Eagle Avenue from the Southern State Parkway. Use that to travel along Lakeside Drive to the southern end.

INFORMATION

http://nysparks.com
https://www.pdf-maps.com/maps/91627

Unless you have a New York State Parks Empire Passport, you'll need to pay a vehicle fee on weekends and holidays during spring and fall and on all days through summer.

Restrooms are in all parking lots and south of the lake at MacDonald Pond. Picnic tables are found throughout. Concession stands operate only in the busy summer season, but you can find vending machines near the restrooms at the first/main parking lot on Lakeside Drive. The toll taker at the lot will let you park for free for fifteen minutes in the little pull-over to the left of the booth so that you can use the facility and buy yourself a snack.

This park used to be visited by thousands of overwintering waterfowl and had been an important refuge. Recent counts indicate dramatically fewer birds, although interesting sightings crop up from time to time.

OTHER THAN BIRDING

Fishing, horseback riding, tennis, basketball, and baseball; dog walking area.

Scope Might be helpful in winter.

Photos Medium difficulty due to often high trees, dense foliage, distance from subject.

Wood Duck among Mallards at Massapequa

Massapequa Preserve and
Tackapausha Museum and Preserve

This lovely 423-acre wooded sliver of green follows a stream and can be enjoyed in any season. There are five lakes in the preserve, and the dams provide nice weedy areas attractive to waterfowl. In addition, its fairly dense forest appeals to migrating and nesting birds. Biking and walking trails will take you the full four-mile length of the park, and some of the trails have great views over open water. The place really comes into its own as a destination during the late fall and early winter when ducks and geese settle on the tranquil series of ponds and lakes fed by Massapequa Creek.

VIEWING SPOTS

You can enter the park through several portals, but you will have the most success birding in the *northern part of the preserve* starting near the Long Island Rail Road station at Massapequa, located off Broadway between the Sunrise Highway and Veterans Boulevard. Once you exit the train or park your car here, enter the park and follow the perimeter path of the first pond that you encounter. You'll continue to head north and will come across two more ponds, at which point you have a couple of choices about which path to

NASSAU COUNTY

195

take. One is the Greenbelt Trail that goes through the woods, and the other is the bicycle path, which is a little better for birding as it follows the stream and circles the ponds more closely.

As you walk north, once you have seen the first three ponds and are making your return, consider a longer hike by passing your starting point at the first pond. Cross busy Sunrise Highway and follow the bikeway through the birdy woods and along the stream that leads to *Massapequa Lake*. You'll walk six miles on this three-pond-on-top, one-lake-on-bottom venture. If you'd like to skip the top part of Massapequa Preserve, you can park south of Sunrise on Ocean Avenue and access the lake and woods from there. However, if you don't have time to do both the upper and lower parts of the preserve, opt for a trip around at least one of the ponds north of the train station, as the woods there are birdier, and you'll have more opportunities to see a wider range of species, including Eastern Screech Owl, Great Horned Owl, Wilson's Snipe, and a variety of migrants.

Close by to the west is *Tackapausha Museum and Preserve*, a narrow park of eighty-four acres accessed by a five-mile trail system that takes you past ponds, wetlands, and more pretty streams—in this case, the interweaving channels of Seaford Creek. It's worth checking out.

KEY SPECIES BY SEASON AT MASSAPEQUA PRESERVE

Spring Songbird migration includes several species of vireos and warblers; Northern Rough-winged Swallows, Blue-gray Gnatcatcher, Eastern Towhee, a few shorebirds, Osprey. Migrating waterfowl leave in April.

Summer Woodland nesting birds like Warbling and Red-eyed Vireos, Wood Thrush, Cedar Waxwing, Baltimore Oriole, Great Crested Flycatcher, Eastern Kingbird; Green Heron, Forster's Tern, Barn Swallow.

Fall Waterfowl arrive and include Northern Shoveler, Gadwall, Pied-billed Grebe; yellowlegs, Great Blue Heron; Cooper's Hawk; Hermit Thrush; Common Yellowthroat.

Winter Swans, Canada Geese, and Mallards are abundant, but this is also a good opportunity to get more interesting ducks like Gadwall, scaup, Ruddy Duck, Common and Hooded Merganser, Ring-necked, and Wood Duck. Massapequa sometimes has Eurasian Wigeon, Northern Pintail, and Redhead. Other overwintering birds include Red-bellied and Downy Woodpeckers; Black-capped Chickadee; Fox and Swamp Sparrows; Eastern Screech and

Great Horned Owls; Long-billed Dowitcher; Wilson's Snipe; Winter Wren; Rusty Blackbird.

HABITAT

Freshwater, forest, edge, salt marsh.

BEST TIME TO GO

Fall and into winter is great for waterfowl until the ponds freeze. There is nice birding throughout the year, although birders like it mostly fall through spring.

HOW TO GET TO MASSAPEQUA PRESERVE

By Car Set your GPS for the corner of Broadway and Veterans Boulevard in Massapequa, New York. If you need a street address, 20 Broadway will get you relatively close.

If you arrive after commuters have left for work, you will probably need to park in the lot at the end of Veterans Boulevard, which is where our tour starts. If that lot is also full, look for legal street parking.

By Train The Long Island Rail Road (LIRR) will take you to Massapequa from Manhattan in about an hour. The train station is very close to the preserve entrance recommended in this tour.

By Bus The N55, N80, and N81 stop at Veterans Boulevard; the N54 stops at the Massapequa LIRR train station.

INFORMATION

Massapequa Preserve
Broadway and Veterans Boulevard, Massapequa, NY
516-572-0200
www.co.nassau.ny.us

Tackapausha Museum and Preserve
Washington Avenue between Merrick Road and Sunrise Highway,
  Seaford, NY
516-571-7443
http://www.discoverlongisland.com
http://fdale.net/FMP

There is a fee for entrance to the museum.

No restrooms or café. The closest restroom is in the Massapequa LIRR train station ticket office lobby at Sunrise Highway (Route 27), just east of Broadway and Route 107.

Train station ticket office hours are limited.

John F. Kennedy Memorial Wildlife Sanctuary and Tobay Beach

Just east of Jones Beach, this 525-acre wildlife sanctuary is located adjacent to the parking lot at Tobay Beach on Ocean Parkway, Massapequa. An important safe refuge for migratory waterfowl and shorebirds, it has a reputation for decent numbers of Long-tailed Ducks and Bufflehead and attracts other more common waterfowl. Guggenheim Pond is the centerpiece of this sanctuary, and before Hurricane Sandy washed out the bridge to it, you could look here for Glossy Ibis and other waders spring through fall and freshwater ducks in winter. The landscape features creeks, dunes, salt marshes, and both freshwater and saltwater ponds. It's not always open, and, sadly, the observation tower is not usable. It might be worth a look if you're already birding in the area, but it's remote, and we don't recommend exploring here alone. If do you plan to give it a try, check first to make sure it's open and that you can get a permit. If it's not, you can always take the path from the parking lot to the beach, make a left, and follow the path that enters and loops around a wild area. These are the outer edges of the sanctuary's wetlands, and while it is not the entire experience, it will give you some views over the refuge.

Enter the Tobay Beach parking lot and drive around to the southwest corner to find the road to the JFKMWS parking lot. Anyone can visit, but you need to get a permit from the town. This may be pricey if you are not a resident of Oyster Bay. Call Parks Department Beach Division at 516-797-4110.

Tobay Beach is a quick fall, winter, and spring waterfowl stop. Look for sparrows in fall. Off-season you won't need a parking permit.

If you continue east, you can make a stop at *Gilgo Beach* and *Captree State Park*, which are described in the Suffolk County chapter.

SOUTH SHORE WINTER FRESHWATER BIRDING

If you're hankering for freshwater ducks in winter, head for Long Island's South Shore. From November through March, when there is open water, you can get good views of a variety of waterfowl—some at fairly close range—on a series of small ponds. Making the rounds, you can reasonably expect to see Common Goldeneye, American and Eurasian Wigeon, Northern Pintail, Ring-necked Duck, Northern Shoveler, Green-winged Teal, Wood Duck, Bufflehead, Pied-billed Grebe, scaup, Ruddy Duck, Redhead, American Coot, and Gadwall. Hooded, Common, and Red-breasted Mergansers may be found among the ubiquitous American Black Duck and Mallard.

Many of these ponds were designed for drainage and lie between Merrick Road and Route 27.

There are a couple of locations in this group that deserve special mention. For whatever reason, *Grant Park Pond* usually has the most interesting assortment of ducks, and since it's a relatively small pond, the birds are close enough to enjoy without a scope, although you would not regret bringing one. *Willow Pond* is truly a find, with a fair amount of ducks and a heron roost, and if you haven't been here, it's worth battling the one-way streets to see it. *Camman's Pond*, another spot that gets little attention, is great for interesting ducks and good photo ops. The larger parks such as *Cow Meadow Preserve* and *Massapequa* have more year-round appeal, so include them in your winter waterfowl tour, but don't overlook them in other seasons.

We list these destinations in order of travel from west to east starting at Grant Park Pond and ending at Massapequa, but you can travel either direction. For this outing you will need a car. These lakes are easy to access, and you could visit all of them in a single day if you were fairly systematic about your walk-arounds. However, that wouldn't give you a lot of time at each spot, and a few of these stops might tempt you to linger. Many of these places don't get a lot of birdwatchers but can be surprisingly productive.

Here is our suggested route, followed by an overview of each location—the best spots are marked with an asterisk.

Grant Park Pond and Willow Pond*

Grant Park Pond and Willow Pond in Hewlett are two scenic freshwater lakes fairly close to each other. Grant Park Pond is part of a public park and is actively fished, while Willow Pond is a small freshwater pond in an intimate setting. It attracts waterfowl in winter and has a heron roost.

From the *Grant Park Pond* parking lot, head straight for the pond in front of you. There is an obvious viewing area from which you can see most of the waterfowl in the pond. Some birds will be fairly close, but like many places, the more interesting ducks are generally farther away, so make sure to check out what is happening at the far end of the pond. The entrance is at Broadway and Sheridan Avenue. Once you are finished at Grant Park, make your next stop Willow Pond.

Willow Pond is a peaceful and beautiful little-known pond that can have a lot of waterfowl. Park on tiny Everit Avenue and be sure to keep off private property and out of the active road. This is a quiet residential community with very little on-road parking, so please be respectful of the people living here. On one side of the road is idyllic Willow Pond. The other side of the road has views of Macy's Channel, which is less tranquil, but birds may be there as well. Before you leave, make sure to check the trees to the left of the pond, as there is a heron roost surprisingly close to the road. The pond is located south of Grant Park; make a left off Broadway onto Everit Avenue to visit this lovely spot.

Lofts Pond Park

Lofts Pond Park in Baldwin is an attractive fourteen-acre park with a central freshwater pond bordered by a residential neighborhood. The main open expanse at the entrance on Merrick Road has good views across the lake, and there can be waterfowl close by the shore. You can get better views of the waterfowl floating against the island from the wooden walkway built over the pond, which makes both viewing and taking photos a bit easier. This is a quick stop, and if you are traveling west to east, Milburn Pond will be next.

Milburn Pond

Milburn Pond, in Freeport, sits on eight acres and has a nice half-mile paved walk around it. The pond is freshwater, while Milburn Creek Park across Merrick Road is brackish. In winter, Milburn Pond is an easy place to see waterfowl.

Waterfowl congregate toward the middle of the pond, but a walk around the entire body of water should provide good close-up views of ducks taking advantage of the less-trafficked area near the boat launch. While making the half-mile circuit, check the ground and bushes for sparrows or other

White-crowned Sparrow

passerines. Once at the concrete launch, you can get better views of teal, mergansers, and the other waterfowl. Your next stop is *Cow Meadow Preserve,* a much larger area and a year-round birding attraction.

Cow Meadow Park and Preserve*

See our full description earlier in this chapter.

Camman's Pond Park*

Camman's Pond Park (also spelled "Cammanns") is a spot we highly recommend for winter waterfowl. This charming eight-acre park with a winding pond and pathway is in a residential area in the town of Merrick. It doesn't get a lot of attention, but it does get a lot of waterfowl. It's a pleasant place to swing by for fairly close-up views of ducks in season and can be a satisfying coda to a day spent birding Long Island's South Shore or Jones Beach.

Camman's Pond Park is located at Merrick Road and Lindenmere Drive. Park on Lindenmere and pick up the path at any point. The paved sidewalk

NASSAU COUNTY

201

Northern Pintail

on the eastern side of the pond runs for approximately one-half mile and has benches. Make sure to check out the island at the north end of the pond, where ducks congregate and where Black-crowned Night-Heron are found. Many birds are within range for easy photography, but if you want the best light, plan to visit in the morning.

Mill Pond Park and Twin Lakes Preserve, Including Wantagh Pond and Seaman Pond

Mill Pond Park is an attractive fifty-four-acre park with hiking trails surrounding a freshwater pond. Just north of Mill Pond Park is Twin Lakes Park, a series of five ponds including Wantagh and Seaman Ponds.

For *Mill Pond Park*, park on busy Merrick Road. Look for waterfowl on the U-shaped pond and especially near the center island. Take a walk around the pond and enjoy a variety of views of the waterfowl. The north part of the park has surprisingly wild woodland trails, and in winter you may find a variety of woodpeckers. You'll see migrants and nesting songbirds the rest of the year. If you plan to photograph waterfowl, morning gives the best light across the pond.

Twin Lakes are fenced off and less accessible, although there is a handy viewing spot from the bridge between *Seaman* and *Wantagh Ponds* and, from there, the opportunity to walk around Wantagh for closer views of waterfowl.

Massapequa Preserve and Tackapausha Preserve

Check out our full description of these hot spots earlier in this chapter.

THE NORTH SHORE

Nassau County's North Shore, known as "the Gold Coast", was a playland for the wealthy in the early twentieth century. Many of the mansions have since been demolished, but its appeal remains, with lovely beaches surrounded by bluffs and the opportunity to bird some of the remaining forests and swamps. The region also has lots of shoreline for winter waterfowl viewing, as well as attractive parks and converted estates.

Muttontown Preserve

At 550 acres, this is Nassau County's largest preserve and a lovely location to bird and explore. Originally three separate estates (including the now-crumbling remains of the estate of King Zog of Albania), they are now combined into one park and are all being allowed to return to a wilder and more natural state. Large swaths of fairly dense woods with winding, easy-to-walk trails open up to large fields and pass into the occasional marshy area. This provides inviting bird habitat for a wide variety of species. During spring migration and early summer it can be especially rewarding. Muttontown is an Important Bird Area and a wonderful spot, not often visited, where you can have a quiet and satisfying birding experience. It's definitely worth a visit if you plan to bird in this area. The trail system is complicated, though, and a GPS and buddy are essential to the first-time visitor.

VIEWING SPOTS

From the parking lot, find the Nature Center, where maps are available behind the building, and take any trail. The trails from the Nature Center take you south through the woods, with places along the main path that can be very birdy—especially as you get closer to the middle of the preserve. The ones that take you east and south around the kettle ponds will be most productive if the ponds have water in them. You may spot a Great Blue Heron. Some paths dead-end into open fields for views of field-loving birds such as Tree

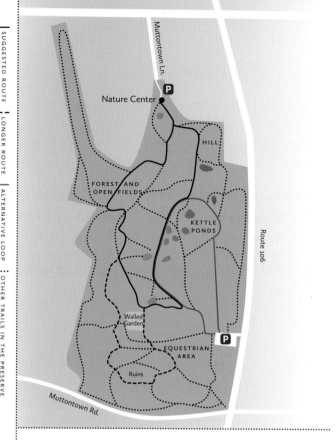

MUTTONTOWN PRESERVE

and Barn Swallows, Common Yellowthroat, and Indigo Buntings in spring and summer. During peak season you may find yourself birding by ear, as the foliage is quite dense.

If you want to go as far as the area around the equestrian center, it's a beautiful hike, but you'll likely find only common backyard birds there. In this section, the terrain is slightly different and a little hillier. Here you will also see greater evidence of the former estates, such as the Walled Garden.

KEY SPECIES BY SEASON

Spring Two dozen species of warbler can be seen here, plus Indigo Bunting, Rose-breasted Grosbeak; Red-eyed, Warbling, and White-eyed Vireos; Veery, Wood Thrush, Tree and Barn Swallows, Song and Chipping Sparrows.

Summer A nesting bird bonanza—Baltimore Oriole, Red-winged Blackbird, Scarlet Tanager, Carolina Wren, Common Yellowthroat, Yellow Warbler,

Blue-gray Gnatcatcher, American Redstart, and a variety of woodpeckers; Great Blue Heron.

Fall Returning songbirds, raptors.

Winter Dark-eyed Junco, Black-capped Chickadee, Tufted Titmouse, Northern Flicker, American Robin, Northern Cardinal, and Blue Jay are year-round residents.

HABITAT

Mature deciduous forest, open wildflower fields, and some marshy bits.

BEST TIME TO GO

Spring for migrants and early summer for nesting birds.

HOW TO GET TO MUTTONTOWN PRESERVE

This preserve is best reached by car. Set your GPS for the Nature Center entrance on Muttontown Lane (no street address), East Norwich.

INFORMATION

Muttontown Lane in East Norwich (south side of 25A)
516-571-8500
http://www.nassaucountyny.gov

The preserve is open daily, and entrance is free. However, the Bill Patterson Nature Center at the site, which has restrooms and maps, is not open during all the hours of the preserve, and when the center is not open, the restrooms are locked. Maps are available at the Nature Center when it is open or are left outside on the back porch when closed.

Parking is at the main entrance off Muttontown Road, or by the equestrian center.

There is no restaurant or café in the preserve, but you can easily find places to eat in the surrounding neighborhood.

It is essential to take a map with you for this preserve. The trails are not well marked, so having a GPS will also help. When the trails are marked, they are fairly easy to follow, but you don't want to wander aimlessly, which is easy to do here.

There is an equestrian center at one of the entrances to the preserve, and many of the unmarked but interesting trails are bridle paths. You may see the occasional horse and rider.

Golden-crowned Kinglet

Because of its beautiful mature forests, this preserve is especially cool and comfortable in the warmer months. It also is relatively quiet and peaceful.

Be aware that while the trails are fairly well groomed, there can be a lot of poison ivy, both alongside the wider trails and in the middle of those that are less used. Much of the preserve is thickly wooded and can be moist. Prepare for ticks and mosquitoes.

Scope Leave it at home.

Photos Challenging when the trees are leafed out.

Leeds Pond Preserve

This tranquil, hilly, wooded preserve overlooks Manhasset Bay and is home to the Science Museum of Long Island. To see birds, take the trails that wind through the woodland and past the ruins of the old farmhouse. Also check out the stream that runs into the pond. Look for waterfowl fall through spring and waders like Great Egret and Black-crowned Night-Heron spring through fall. The woods are attractive to sparrows, warblers, and other songbirds like Rose-breasted Grosbeak. Find Leeds Pond Preserve at 1526 North Plandome Road, Plandome Manor; 516-627-9400.

Sands Point Preserve

Three mansions, including a former Guggenheim "castle," are now part of a 216-acre preserve of manicured gardens, forests, meadows, a freshwater pond, cliffs, and beach along Hempstead Bay. Back in the day, F. Scott Fitzgerald found material here for some of the indulgences in *The Great Gatsby*, and you can indulge your desire for birds while visiting these former Gold Coast estates.

VIEWING SPOTS

There are six trails that lead to various parts of the preserve, but you'll want to head to the pond, the forested area, and the beach. In spring, over two dozen species of warbler are found in the woods; look for Osprey, Spotted Sandpiper, Red-eyed Vireo, and Chimney Swifts in summer; Cedar Waxwings, Eastern Towhee, and other sparrows in fall; winter waterfowl include Red-breasted Merganser, Long-tailed Ducks, Red-throated Loon, and the occasional Barrow's Goldeneye, plus gatherings of Ruddy Turnstones. Sands Point is a similar experience to Caumsett State Park in Suffolk County, only smaller and not as birdy.

HOW TO GET TO SANDS POINT PRESERVE

Reachable only by car. There is a per vehicle entrance fee in all seasons. Use Exit 36, Searingtown Road, from the Long Island Expressway (I-495) and take Searingtown Road north as it changes names, finally becoming Middle Neck Road.

INFORMATION

127 Middle Neck Road, Sands Point, NY 11050
516-571-7900
www.TheSandsPointPreserve.com

Restrooms are open year-round in the visitor center, where you can also buy drinks and snacks from the vending machines.

Whitney Pond Park

Twenty-four acres of pretty woodlands have as their main natural attraction Whitney Pond (sometimes called Whitney Lake). Walk the paths and try the overlook onto the water, where you may find wintering waterfowl. Spring through fall brings Wood Ducks, Osprey, waders, and migrating shorebirds. Check the trees for migrating warblers and for woodpeckers year-round. On Community Drive at Northern Boulevard, Manhasset; 516-571-8300.

William Cullen Bryant Preserve

Get a boost of culture and elegance along with your birds at this preserve, home to the Nassau County Museum of Art. Formal gardens and woods grace 141 acres. Woodpeckers, nuthatches, and chickadees year-round; warblers pass through spring and fall. Listen for Carolina Wren. Located at 1 Museum Drive, Roslyn Harbor.

Garvies Point Preserve

This attractive sixty-two-acre park has five miles of wide, well-groomed trails with beautiful views of Hempstead Harbor. Look for Baltimore Orioles in the parking lot. Head for the fields and check the edges for warblers. The cliff trail is woodsy and has the great views, but be sure to check the shoreline first for activity before making the descent to the beach. The forests can also be fairly productive. Garvies Point is a small, pleasant place to visit spring through fall. The entrance and parking lot are at 50 Barry Drive, Glen Cove, NY 11542; 516-571-8011.

Welwyn Preserve

Walk any of the four nature trails in this 204-acre preserve converted from a former estate. The grounds were designed by the Olmsted family, and the house is now the Holocaust Memorial and Tolerance Center of Nassau County. There are a number of habitats to explore, including dense woodlands, freshwater ponds and swamps, plus a salt marsh and stretch of sandy beach on Long Island Sound. The path to the left of the house is an easy downhill walk to the beach, whereas the trails behind and to the right of the house are a little more difficult. In the woodlands look for migrating songbirds, fall and winter sparrows, and woodpeckers; the shore should provide some shorebirds and Osprey, and there are egrets in the salt marsh.

It might not be the birdiest trip, but you will have the chance to see a lot of habitats—and with diversity there is potential! Crescent Beach Road, Glen Cove, NY; 516-571-8040.

Stehli Beach Preserve and Charles E. Ransom Beach

Stop at these neighboring Bayville beaches in winter for waterfowl sightings. Both offer great looks at Long Island Sound, but Ransom Beach does so in particular, with its elevated views from a raised parking lot. Overwintering waterfowl can be seen in decent numbers from the beach. Look for Long-tailed Duck, Common Goldeneye, Red-breasted Merganser, Common Loon, and Horned Grebe. Occasionally Barrow's Goldeneye make an appearance. Migrating scoters turn up, sometimes in flocks; fall has some shorebirds. Late November through March is best. Stehli Beach may also host nesting Least Terns in summer, and check the marshy areas for herons and egrets. Stehli Beach parking is at the intersection of Oak Neck Beach Road and Bayville Avenue. Ransom Beach parking is just east on Bayville Avenue.

Centre Island Town Park*

A terrific winter birding stop on Oyster Bay for waterfowl, including Common Goldeneye, Red-breasted Merganser, Horned Grebe, scaup, and scoters. Park either in the Centre Island Beach Villagepark, or stop on Bayville Road to view birds in the waters on both sides of the neck. Be aware that hunters also know about this location, and if you don't want to see Long-tailed Ducks being shot as you watch them fly past, you might want to avoid it during hunting season. Also a great place to scan the ground for Horned Lark and Snow Bunting.

Bailey Arboretum

This former estate of an amateur botanist has lovely gardens and its own Wildlife Rehabilitation Center. Visit here in spring for pretty flowers and a smattering of songbirds, and in fall for sparrows. If you can't find birds in the gardens, you can stop in at the rehab center, which often has a variety of rescued owls and other raptors. Located at 194 Bayville Road, Locust Valley; 516-801-1458; http://www.baileyarboretum.org.

Shu Swamp (Charles T. Church Nature Sanctuary)*

"Shu" is an old Dutch word meaning "cascading waters," and there is freshwater aplenty here in the form of wetlands, gurgling streams, and Beaver Brook. This lovely preserve comes alive with flowers and birds in spring. Visit the swampy areas and mudflats for shorebirds. Check the woods for Wood Thrush, spring warblers, nesting Great Horned Owls, and Wood Ducks. Take the trail south to the overlook onto Lower Francis Pond (see below). The sanctuary is in Mill Neck, on Frost Mill Road just south of the Long Island Rail Road overpass.

Upper and Lower Francis Ponds

Near Shu Swamp, across Frost Mill Road on Beaver Brook Road, you can find these two ponds, good for winter ducks. The best way to visit is to walk over from Shu Swamp.

Planting Fields Arboretum State Historic Park

Planting Fields in Oyster Bay is a lovely 409-acre former Gold Coast estate boasting a serious horticultural program and a Tudor-revival mansion with original furnishings. It's best for songbirds during spring and summer, and because it's so charming, locals prefer birding here to some of the other preserves. If you don't have an Empire State Pass, you'll pay to park from April to November, but where else are you going to find a greenhouse devoted strictly to camellias? Planting Fields has restrooms, vending machines for snacks and drinks, and a café that may have limited hours. Find it at 1395 Planting Fields Road; 516-922-9200. It is one and a half miles from the Oyster Bay stop on the Long Island Rail Road.

Mill Pond in Oyster Bay

If you are in the area, Mill Pond is an easily accessible winter-only stop for waterfowl. At West Main Street and Lake Avenue.

Sagamore Hill

This National Historic Site was Theodore Roosevelt's home and was known as the Summer White House during his presidency. Roosevelt's love of nature is evident here as the site has a variety of habitats and a bird list of over 115 species. The eastern portion of the property has larger parcels of natural land, including mature forest, salt marsh, and beach. Take the nature trail through these habitats to the shoreline of Cold Spring Harbor for winter waterfowl, including Long-tailed Ducks; look for egrets and herons in the salt marshes in summer. With pastureland, open fields, and orchards, there is the chance to see Eastern Bluebirds as well as Red-winged Blackbirds and sparrows spring through fall. The preserve also has Winter Wren, Great Crested Flycatcher, Yellow-billed Cuckoo, and Ruby-crowned Kinglet. Twenty-six bird species are confirmed breeders on the property, with another two dozen species probable nesters. Sagamore makes for a nice outing to see a piece of history and the birds that live on the property. Located at 20 Sagamore Hill Road off Cove Neck Road, Cove Neck; 516-922-4788; www.nps.gov/sahi (also provides a bird checklist).

St. John's Pond Preserve

St. John's Pond Preserve is part of a Nature Conservancy complex in Cold Spring Harbor. Its reputation is hit or miss, but when it's a hit you'll find Eurasian Wigeon, Ring-necked Ducks, and Redheads; Tufted Duck have also been seen here. Park in the Fish Hatchery and Aquarium lot and use the stairs to look out over the pond. While you are there, take advantage of the location and walk along the paved walkway at St. John's Church for another view of the pond from the west side. Spring and summer are less interesting but still worth a visit. In fact, you can also drive to and park in the laboratory parking lot across busy Highway 25A and follow the road to the end of the spit of land to see nesting Osprey. There are a couple of trails with views of the harbor, and while they might not be super-birdy, they offer a beautiful walk spring through fall to see shorebirds and even the occasional Bald Eagle. If you have the time, prolong your visit by strolling around in some of the oldest forest on Long Island. The preserve is only fourteen acres total, and close enough to combine with a trip to *Uplands Farm Sanctuary*. For St. John's, park at the fish hatchery and get a key to the gate if you want to access trails other than the one to the pond; 1660 Highway 25A, Cold Spring Harbor.

Uplands Farm Sanctuary

Uplands Farm Sanctuary in Cold Spring Harbor is a Nature Conservancy preserve with woodlands and meadows and a well-maintained trail system. Stop in at this once-working farm during spring and fall migration. Blue-winged Warblers can be seen here, and it's one of the few places in Nassau County to see Bobolink. Eastern Bluebirds return every spring. If you are on a Long Island Greenbelt hike, this sanctuary connects to the Nassau-Suffolk Trail. Parking is reached off Lawrence Hill Road; 631-367-3225.

7 Suffolk County

This part of Long Island was originally inhabited by the Algonquin Native Americans, who used the area's rich marine environment for harvesting shellfish. Whereas most of the New York City area was settled by the Dutch, Suffolk County was colonized by the English, who named it after the county from which its first settlers came in the seventeenth century. In time it became important for whaling and farming.

Suffolk's strength as a birding locale lies in its geographical position, reaching 120 miles into the Atlantic, and its wide variety of habitats, which include grasslands and pine barrens along with maritime forests, wetlands, and beaches. As you travel away from New York City, the dense suburban settlement of central Long Island begins to give way to wilder settings, and the concentrated songbird migration of the other counties becomes somewhat diffuse. However, no other place in southeastern New York provides the numbers and species of seabirds, shorebirds, and pelagics that can be found here. This means that any time of year there will be some kind of migrating bird that can be found in Suffolk County.

Because Suffolk is so huge in relation to the other counties covered in this book, and because, compared to the others, it is relatively sparsely settled, there is an almost endless number of places to stop and look for birds. We have selected the most interesting ones. In your travels, you will likely come across local birders who will be more than happy to tell you about their favorite spot. To get you started checking out this vast area, we offer some highlights.

Suffolk County has two of the three top birding hot spots on Long Island—*Montauk Point* and *Robert Moses State Park*. If you go nowhere else in Suffolk, you must see these. Montauk Point is good for birds year-round, but is known for its awesome numbers and diversity of waterfowl, pelagics, and seabirds. Open-water pelagic tours leave from Montauk as well. Robert Moses State Park has a fall hawk watch, but its westernmost wild area, Democrat Point,

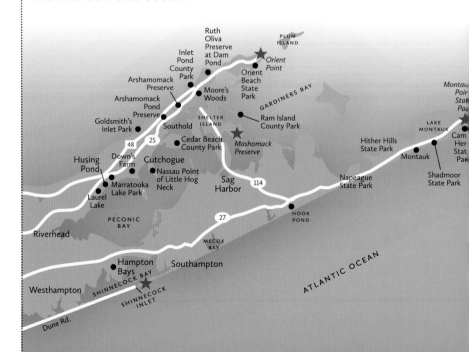

WESTERN SUFFOLK COUNTY

CONNECTICUT

59

Bridgeport

15

Norwalk

95

LONG ISLAND SOUND

Stamford

Wading
River Marsh
Preserve

Golde
Triang
sc
farm

Wildwood
State Park

Caumsett
State
Historic
Park

Target
Rock
NWR

Sunken
Meadow
State
Park

Hulse
Landing Rd.

Buffa
Farm

Tung Ting
Pond and
Mill Pond

Stony Brook

EPCAL

David Weld
Sanctuary

Prestons Pond
Calverton Ponds Preserve
Swan Pond

110

25

495

MANORVILLE

Syosset

Blydenburgh
Park

Shirley

Connetquot River
State Park

27

Patchogue

Wertheim
NWR

Cupsogue
Beach
County
Park

Bethpage

Islip

Bayard Cutting
Arboretum

Smith
Point
County
Park

Heckscher
State Park

Watch
Hill

Freeport

Captree
State Park

Sunken
Forest
at Sailors
Haven

Robert Moses
State Park

Gilgo
Beach

Jones Beach
State Park

EASTERN SUFFOLK COUNTY

Ruth
Oliva
Preserve
at Dam
Pond

PLUM
ISLAND

Inlet
Pond
County
Park

Orient
Point

Arshamomack
Preserve

Moore's
Woods

Orient
Beach
State
Park

GARDINERS BAY

Montau
Poir
Stat
Pa

Arshamomack
Pond
Preserve

SHELTER
ISLAND

Ram Island
County Park

LAKE
MONTAUK

Goldsmith's
Inlet Park

Southold

Cedar Beach
County Park

Mashomack
Preserve

Hither Hills
State Park

Cam
Her
Stat
Pa

Husing
Pond

48

25

Down's
Farm

Cutchogue

Sag
Harbor

114

Montauk

Shadmoor
State Park

Nassau Point
of Little Hog
Neck

Napeague
State Park

Marratooka
Lake Park

Laurel
Lake

PECONIC
BAY

27

HOOK
POND

Riverhead

MECOX
BAY

Hampton
Bays

Southampton

Westhampton

SHINNECOCK BAY

ATLANTIC OCEAN

SHINNECOCK
INLET

Dune Rd.

is active year-round. A trip to *EPCAL*'s former airport and *the sod farms* area will take you to a very special pine barrens and grassland environment. Visit *Sunken Meadow* year-round, but maybe avoid the crowds by staying away in summer. For guaranteed quality winter waterfowl, stop in at *Shinnecock Inlet.* It's the first stop along *Dune Road*, a pull-over route that will take you to a variety of good viewing spots. For something a little different, try *Caumsett,* a gorgeous former private estate with more habitats and property to explore than is possible in one day. And for a trip back in time, to what the Hamptons were like when they were sleepy little towns along the Atlantic, visit Shelter Island and its magnificent preserve, *Mashomack.*

You will absolutely need a car to visit most of these places, although a couple of them are serviced by train. Day trips are possible, or combine a few sites together for an unforgettable weekend in any season.

WESTERN SUFFOLK

Robert Moses State Park

One of the top hot spots in Suffolk County and designated an Important Bird Area, Robert Moses State Park is located at the far western end of Fire Island. With five miles of beaches on the Atlantic Ocean, it's also popular with beachgoers and is visited by nearly four million people each year. The land was originally purchased in the nineteenth century for a lighthouse and was dedicated as a park in 1908. Birders come mostly in the fall during its famous hawk watch, although you can use the trip out to roam around and find a number of songbird migrants. In summer it plays host to a number of beach-nesting birds, including some endangered species, and under the right conditions it can have a productive sea watch.

VIEWING SPOTS

For the fall *hawk watch*, park in Field 5, the easternmost of the lots, and walk east along the boardwalk. The watch is considered one of the best on Long Island, and it's manned by members of the Fire Island Raptor Enumerators. Try to time your visit on a day from mid-September to mid-October—early October is prime—when north or northwest winds can bring hundreds of raptors streaming overhead in a single day.

This is also a good time to continue around Robert Moses and look for

Common Eiders

American Oystercatcher and chick

other fall migrants. Use the boardwalk to walk through the brush around the *lighthouse* and to visit the waters of the Great South Bay. From here, you could continue your walk along the dirt road toward the village of Kismet, but if you're pressed for time, return to your car and drive west to *Parking Field 2*. Bird the area around here and you may get some more interesting fall migrants: Clay-colored Sparrows, Dickcissel, and Bobolink are sometimes found. Fall butterfly migration is an added bonus.

Spring has some nice songbirds, and a sea watch is often productive. Summer brings more people than birds, but by August shorebird numbers are really increasing, so visit on a weekday to avoid weekend crowds. In winter, waterfowl is best seen from the beach. Use Parking Field 2, which is the same lot you'll use for a summer sea watch.

Democrat Point is the westernmost tip of Fire Island, and because it is relatively difficult to reach, it doesn't get a lot of birders. It's more of a year-round location where you can see seabirds and nesting shorebirds like Piping Plover and American Oystercatcher late spring through summer. A trek out there during the spring and fall may net you some interesting songbirds, as well as Merlin and Peregrine Falcon. For shorebirds, time your visit during low tide, when they'll be feeding on the mudflats. Fishermen with permits may drive out to the jetty, nearer the tip. The rest of us have to park in Field 2 and trek out past the pitch-and-putt golf course (which can have interesting grassland birds), a distance of about one mile over dirt and sand.

KEY SPECIES BY SEASON

Spring Kinglets, Thrushes, some warbler activity, but it's better in the fall. Brown Thrasher; Orchard and Baltimore Orioles (also around in fall); Sparrows such as Chipping, Field, and Swamp; Blue Grosbeak and Indigo Bunting. Eastern Meadowlark sometimes appear.

Summer Great Blue Heron, Great Egret, Snowy Egret, Osprey. Uncommonly, pelagics will make an appearance, such as Sooty and Cory's Shearwater, Wilson's Storm-petrel, and Parasitic Jaeger. American Oystercatcher, Piping and Semipalmated Plovers, Killdeer, Willet, and other shorebirds; Laughing Gull and sometimes Lesser Black-backed Gull; Terns including Royal, Roseate, and Black; Willow Flycatcher, Eastern Kingbird; Tree, Bank, and Barn Swallows; Boat-tailed Grackle. By August, migrating shorebirds will include Least and Semipalmated Sandpipers and, as the season progresses, occasional Baird's, Upland, Pectoral, and Western Sandpipers.

ROBERT MOSES STATE PARK

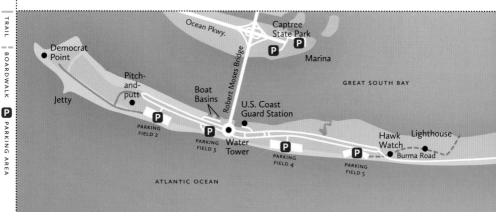

Fall Raptors include Northern Harrier, Sharp-shinned and Cooper's Hawks, Peregrine Falcon, Merlin, and American Kestrel; some Bald Eagles. A wonderful time for sparrows, and you may get a Clay-colored, Vesper, or Lark; Blue Grosbeak, Indigo Bunting, and occasionally Dickcissel, Bobolink, and Eastern Meadowlark; Vireos, of which Blue-headed and Red-eyed are the most common; Swallows may include Cliff; migrating warblers; Northern Gannets are present through spring.

Winter Scoters, Long-tailed Ducks, Common Goldeneye, Red-throated and Common Loon, Horned Grebe. Gulls are abundant, and you may find a Bonaparte's or Iceland. Horned Lark, American Pipit, Lapland Longspur, and Snow Bunting are found. Downy Woodpecker, Carolina Wren, American Tree Sparrow, Purple Finch. Savannah Sparrows could include the Ipswich subspecies. In irruption years, Red and White-winged Crossbill, Common Redpoll, and Pine Siskin are possible.

HOW TO GET TO ROBERT MOSES STATE PARK

At the foot of the Robert Moses State Parkway, use the water tower in the traffic circle to orient yourself around the park. The hawk watch and light-house are on your left, Parking Field 2 and the trailhead to Democrat Point are on your right.

INFORMATION

Robert Moses State Parkway, Babylon, NY 11702
631-669-0449
http://nysparks.com/parks
Fire Island Raptor Enumerators: http://www.battaly.com/fire
For information on tides: http://tides.mobilegeographics.com and search for Democrat Point
There are lots of ticks in season, so dress accordingly.

OTHER THAN BIRDING

Anything beach related, including swimming and fishing; there is a pitch-and-putt golf course, volleyball courts, picnic areas. Surfing is allowed after Labor Day.

Scope Bring one if you're looking for shorebirds.

Photos Bring a long lens.

Sunken Forest at Sailors Haven, and Watch Hill

East of Robert Moses are two remote and beautiful Fire Island preserves that bear mentioning. Even though, as the crow flies, they are quite close, you need to plan in advance to visit either of these sites. They cannot be reached by simply driving eastward, but have limited access by water and on foot.

The *Sunken Forest at Sailors Haven* is a fifty-acre mature forest that provides dense cover and is a good food source for migrating warblers, sparrows, and other forest-dwelling birds. This rare maritime holly forest is found only behind well-established sand dunes along the mid-Atlantic coast. The 1.5-mile boardwalk also takes you through freshwater bogs. Sunken Forest is accessible only by ferry, private boat, and on foot, and is open only from mid-May to mid-October. Check for day trips at http://web.mta.info/lirr/getaways/beach/sunkenforest.htm.

Sailors Haven Visitor Center: 631-597-6183
http://www.nps.gov/fiis
Sayville Ferry: http://www.sayvilleferry.com

Watch Hill has a tidal marsh on the Great South Bay, where you can find herons and egrets, Glossy Ibis, Common Tern, and other shorebirds, including Piping Plovers, which nest here most years. It is located on the western edge of the Otis Pike Fire Island High Dune Wilderness, directly across the Great South Bay from Patchogue. Like Sailors Haven, Watch Hill is accessible only by foot, private boat, or ferry. If you decide to try it, check for day trips: http://web.mta.info/lirr/getaways.

Watch Hill Visitor Center: 631-597-6455
http://www.nps.gov/fiis
Davis Park Watch Hill ferry: http://www.davisparkferry.com

The Montauk Branch of the Long Island Rail Road will take you to Patchogue, where you can walk to the Watch Hill ferry terminal. If you are going to Sailors Haven, take the Long Island Rail Road to Sayville and a Colonial Taxi to the Sayville ferry. Train schedules for the LIRR can be found at http://web.mta.info/lirr/Timetable.

Dunlin and Sanderlings

Smith Point County Park

This enormous and very popular summer beach park is found on the eastern end of Fire Island. Great shorebirds can be found at this Important Bird Area in the summer, between crowds of people and the off-road vehicle traffic on the beach. But its allure comes in the fall and winter when the crowds have left and you are gloriously alone in this part of the Fire Island National Seashore, sharing the wide white beach with only the birds and fishermen.

VIEWING SPOTS

After crossing the Smith Point Bridge, make a right at the west side of the traffic circle and check in at the *visitor center* in the wooden pagoda-like structure. Stopping here is allowed for fifteen minutes before you have to move the car to the big parking lot. As you drive in there, stop at the *fishing pier* at the northeast corner of the lot and see what might be going on in the bay. You can also access the bay from the bridge, and you'll have another opportunity to do this as you walk through the dunes.

After parking, walk to the *fenced-in lawn* just to the west of the main facility and in between the lot and the beach. In fall, some great sparrows can be lurking here. Make your way back toward the visitor center and take the *boardwalk through the dunes*, looking for Yellow-throated Warblers, kinglets, and Merlin. After about one hundred yards the boardwalk ends, and you can make a right along a natural spillway to look in the bay for Common Goldeneye and other waterfowl. Now make your way to the Atlantic beach and turn west to walk to the Hurricane Sandy–carved *breach at Old Inlet*. As you go, you'll want to keep your eyes on the water for gannets, scoters, Long-tailed Ducks, and on the beach for Sanderling and Dunlin. At the breach, there will be more action, and you may find Royal Terns among the seabirds. Rarities like Parasitic Jaeger, Black-legged Kittiwakes, and other pelagic seabirds have been reported. The breach may close sometime in the future, either by natural processes or human intervention.

Smith Point also has an eastern side, known as the Outer Beach, where bay and ocean birds in fall and winter can be found. People pay a per-season fee to drive on certain beaches, and users expect that these beaches will always be open to vehicles. During beach-nesting season there may be tension between fee payers and those trying to keep the Piping Plovers and other birds safe. The compromise sometimes consists of roads cut through the roped-off nesting areas to accommodate off-road vehicles. This makes for a very dicey breeding area as well as difficult viewing.

While we like Smith Point, you might want to consider Shinnecock as an alternative. At Smith, the hike to the breach is two miles each way (mostly over beach sand), and the birds aren't all that close. For decent views of birds, you will need to lug your scope.

If you do go, consider making a stop at the *Shirley Marina County Park* for migrating sparrows in the fall or anything else that might be interesting at the boat launch on Great South Bay.

BEST TIME TO GO

Fall and winter.

HOW TO GET TO SMITH POINT COUNTY PARK

By Car Use the William Floyd Parkway and take it to its end.
By Bus Service from the town of Shirley is offered only in the summer.

Fire Island, Suffolk County office, Shirley: 631-852-1313
http://www.suffolkcountyny.gov/Departments/Parks/Parks

The restrooms at the visitor center are the only ones that are open in the off-season.

Suffolk County bus schedules: http://www.sct-bus.org/schedules and click on 7D or 7E service.

Robert Angelora's Earth Science Research Project, a Self-Guided Science Walk at Smith Point County Park: pbisotopes.ess.sunysb.edu/esp, and find Smith Point County Park under "Science Walks off campus."

Nearby, the *William Floyd Estate*, located at the end of Neighborhood Road, conducts bird walks a few times a year. The grounds are open even when the house museum is closed. http://www.nps.gov/fiis.

OTHER THAN BIRDING

A popular surf-fishing site; surfing; camping by reservation. In summer there is just about everything you'd expect from Suffolk County's largest beachfront park, including food, live music, playgrounds, showers, baseball, basketball, and, of course, Piping Plovers.

Wertheim National Wildlife Refuge

Wertheim is a 2,550-acre area of protection around the scenic Carmans River, providing habitat for an estimated three hundred species of birds in oak-pine woodlands, grassland, and estuary. This Important Bird Area is open only to passive enjoyment, with the exception of some specified hunting days. Often, large parts of the park are inaccessible. During the summer, the river is full of paddlers and fishermen, putting a damper on birdwatching. Spring and fall are the best times to visit, as there can be a good concentration of songbirds during migration. It's also a great place to look for Wild Turkey and Osprey.

VIEWING SPOTS

There are six miles of hiking trails through this forested park and freshwater wetlands. If you can manage a kayak, this is also an excellent route to take for birding through the marshy area. If you prefer to walk, there are two main trails that wind through this Central Pine Barrens preserve. Both leave from the impressive visitor center. In all seasons, try the Connector Trail to the White Oak Trail first. It has several vistas out onto the Carmans River via the overlooks. At the end, there is a platform that has a spectacular view onto the marsh, but it's a two-mile hike in each direction. Much of the White Oak Trail is wide and packed with gravel or cinders, so while ticks and poison ivy won't be much of a problem, be prepared for biting insects.

Alternatively, you can take a longer hike that only has views of the river, by taking the *Black Tupelo Trail* through the woods to Indian Landing.

Patchogue Lake, west of Wertheim, in Patchogue, is worth a winter stop for waterfowl; you may find Northern Pintail, Eurasian Wigeon, and Ring-necked Ducks.

HABITAT

Woodland, river, estuary, some grassland.

BEST TIME TO GO

Fall and spring.

HOW TO GET TO WERTHEIM NATIONAL WILDLIFE REFUGE

Take Highway 27 (Montauk Highway) to Smith Road in Shirley and turn south or set your GPS for 340 Smith Road, Shirley, NY 11967. The parking lot is located at the visitor center.

INFORMATION

340 Smith Road, Shirley, NY 11967
631-286-0485
http://www.fws.gov

Restrooms are located at the visitor center and across the river at the start of the White Oak Trail.

The preserve is highly managed, and hunting is allowed, usually on certain days during October to January, at which time the park is closed to all other activities, including birdwatching. Parts of the preserve can be inaccessible at other times as well, so it's best to call ahead to see what's going on.

Kayaks, canoes, and paddleboards can be launched at a Department of Conservation site on Highway 27 at the intersection of Smith Road. Boat rentals are available at an independent vendor, Carmans River and Kayak.

Captree State Park and Gilgo Beach

The eastern end of Jones Beach Island isn't as birdy as the western end, but it has a couple of stops that might be worth a look if you have the time. Captree State Park gives harbor to a large fleet of charter fishing boats. Its ocean and bay beaches attract winter ducks, and there are marshy areas that appeal to waders.

Gilgo Beach, west of Captree and east of Jones Beach, offers a beach and marshy area, plus good winter waterfowl. Both these locations are Important Bird Areas.

VIEWING SPOTS

At *Captree State Park*, use the parking lot on the bay side where the charter fishing boats are and look out over the marshes to the north. The *upper level lot* overlooks the ocean and has a terrific view of nesting gulls in April and May. Check the grounds around *Captree's Marina* for sparrows.

From Captree, drive west past Gilgo State Park and make your way to *Gilgo Beach* and the beach parking lot on the bay side. From there you can look across the marshes. At the eastern side of the parking lot, take *Cottage Walk*, where you may see Blue Grosbeak and Indigo Bunting during migration. Gilgo Beach is a private beach that charges hefty fees in summer. It also often has areas that are off-limits during Piping Plover nesting season. The beach itself is a good late-fall and winter waterfowl viewing spot. Bring the scope.

This stretch of road is remote and should not be visited alone. We hesitate to recommend this notorious body-dumping ground at Gilgo Beach, but it is birded, and you will hear about it. If you do go, use caution, and go with a friend.

Captree State Park
3500 East Ocean Parkway, Babylon, NY 11702
631-669-0449
http://nysparks.com

Gilgo Beach
Ocean Parkway, less than a ten-minute drive west of Gilgo State Park,
    Babylon, NY 11702
631-669-0449
Beware of ticks in the scrubby areas.

Caumsett State Historic Park

Named after the Matinecock Indian word for "place by a sharp rock," this was once the private residence and gentleman's estate of Marshall Field III. Stunningly situated on the north shore of Long Island, it is a beautiful and inviting park of 1,750 acres with great secluded views of Long Island Sound, deep woods, and a variety of other habitats. This splendid park hosts over two hundred species of birds, including endangered Piping Plover, Osprey (including a nesting pair on Field's house), songbirds, and waterfowl. It has such a rich diversity that it has been designated part of Audubon New York's Huntington and Northport Bays' Important Bird Area, and it truly lives up to its designation.

Caumsett has an extensive system of hiking trails, as well as bridle paths, beaches, and historic buildings. While this park rates high on the list for species diversity, as well as being a delightful place to visit, if you are interested only in birding, this might not be your first choice. The park is large and requires a lot of walking; plus, there is a fee to park. But for a combined visit to a lovely park with historical significance, as well as a whole lot of birds to keep you busy, it can be an amazing outing. If you prefer to get around by bike, bring one. Exploring the grounds that way could save you a ten-mile hike.

VIEWING SPOTS

A round-trip circle tour of the park highlights the Osprey nest atop the Marshall Field House, amazing views of Long Island Sound, and the shorebirds seen on the beaches, the salt marsh, and open fields.

SUFFOLK COUNTY

225

Starting from the parking lot, go past the dairy and take the main path; then continue to the right as it turns into a service road going past the Winter Cottage and stables, which will be on your left. Take some time to investigate the area around the *stables*. Past the stables, find the *path to your right into the woods*, which becomes mostly grassland. This little path can be very birdy and could include Chipping Sparrows, Baltimore Orioles, and Carolina Wrens, all of which nest here. Continue following this path until it meets a paved path, and make a right. This will take you to the *Masters House*, which was the home of Marshall Field. The house faces open fields, where you can see swallows hunting and a large tree that attracts a variety of birds, including sparrows, catbirds, mockingbirds, and cardinals. Now walk to the *back of the house*, where there are benches and great views of open fields, a freshwater pond, and Long Island Sound. But don't be distracted by this scenic landscape, as arguably the best view is behind you, of the Osprey nest on the chimney of this elegant home. In summer the family of Ospreys will be active and easy to see. Chimney Swifts hunting above them complete the picture.

Walk down the fields toward the *freshwater lake*, past the fragrant flowering bushes, and onto the *beach*. Scan the rocky shoreline for sandpipers and other shorebirds. Cormorants can be seen in the Sound. Make a left and walk along the beach, then, at the point, go inland and take the path that leads up along the *shoreline bluffs*. This path weaves in and out of the woods, and when it opens to the coast, there are more spectacular views of Long Island Sound. Look below you onto the rocks for shorebirds, including Semipalmated Plovers and Sanderlings.

This path will eventually lead you to the fisherman's parking lot at the *beach and salt marsh*. (If your main goal is getting to the beach, you can drive to this lot from the main car park if you have a fisherman's permit, avoiding a long trek on foot.) Walk through the parking lot and take the path that separates the beach from the salt marsh. In the spring and summer, you will see the roped-off area where Piping Plovers nest to the right of the path and the salt marsh to the left. Look for Osprey, egrets, and other waders in this area. Keep an eye out for poison ivy, as it can be abundant.

Once you have taken in the sights at the salt marsh, return to the parking lot and make a right onto Fisherman's Drive. At this point, you have two more miles to go before you get back to the car, but they can be miles well spent.

The road begins with a wooded area and then opens onto a wildflower field, which holds promise for grassland birds like Eastern Bluebird (look for their nesting boxes in the field). Red-winged Blackbird and American Goldfinch can be found here in summer, while the trees hold Baltimore Orioles. It's a charming, pastoral walk, made even more idyllic as you share the road and fields with horses and riders.

Continue on this road and you'll see the dairy on your left and open fields, which may be a nice place to spot a variety of swallows such as Barn, Tree, and Bank. There is a series of trees on the right where sparrows gather. Check in the trees and on the ground around them for Chipping, Song, and other species of sparrow. Make a left here into the dairy complex and continue to the parking lot.

For a shorter outing than the ten-miles described above, simply reverse the course and walk up Fisherman's Drive to the beach. It's not a tour of all the spots, and doesn't include the Osprey nest, but you'll visit open fields, woods, beach, and marsh for less than half the effort.

KEY SPECIES BY SEASON

Spring Many neotropical migrants stop over at Caumsett, including Blue-winged, Yellow, Magnolia, and Black-throated Blue Warblers; Common Yellowthroat, Northern Parula, Ovenbird, Scarlet Tanager, and a variety of other songbirds; Bobolink, Rose-breasted Grosbeak.

Summer Endangered Piping Plovers and Least and Common Terns nest here, as do Osprey, Red-winged Blackbird, Baltimore and Orchard Orioles, Willow Flycatcher, Wood Thrush, Veery, American Goldfinch, Bluebirds, and a variety of other songbirds; shorebirds such as Semipalmated Plovers and Sanderlings; Chimney Swifts, Bank, Tree and Barn Swallows; an abundance of sparrows, including Chipping and Song.

Fall Bald Eagle, Northern Harrier, Sharp-shinned Hawk, Cooper's Hawk, Northern Goshawk, Red-shouldered Hawk, Golden Eagle, Peregrine Falcon, Merlin, and Eastern Screech and Great Horned Owls; Whip-poor-will, Red-headed Woodpecker, Killdeer; Field, Swamp, White-crowned, and Savannah Sparrows; a chance at Vesper Sparrow; migrating shorebirds and returning songbirds.

Winter There is an abundance of Mallards and Black Ducks, but also Common Goldeneye and Long-tailed Ducks.

About two-thirds of Caumsett is forest, predominately oak tulip-tree forest. Other habitats include successional old field, low salt marsh, marine eelgrass meadow, maritime beach, successional shrub land, and salt shrub.

BEST TIME TO GO

There are birds here year-round. Spring through fall might be the most interesting, but winter has abundant and sometimes unusual waterfowl.

HOW TO GET TO CAUMSETT STATE HISTORIC PARK

Take Exit 45 from the Long Island Expressway and take Woodbury Road and West Neck Road to Lloyd Harbor. There is a fee to enter. Caumsett is not served by public transportation.

INFORMATION

25 Lloyd Harbor Road, Huntington, NY
631-423-2770
http://www.dec.ny.gov/outdoor
http://nysparks.com
http://www.stateparks.com/caumsett.html

There are restrooms at the entrance in the dairy and at various buildings within the park. No restaurant, but there is a vending machine in the dairy area near the restrooms. Benches can be found throughout the park.

Be alert for ticks. There are signs posted throughout the park warning about ticks, so dress appropriately and avoid tall grass.

Scope Yes—would be helpful along the shoreline and in the salt marsh, but consider how many miles you really want to haul it.

Photos Easy.

Target Rock National Wildlife Refuge

This gorgeous eighty-acre park on the east end of the peninsula shared by Caumsett is part of the Long Island National Wildlife Refuge Complex. It's similar to Caumsett in its diversity and spectacular views, but its much smaller scale makes it easier to visit. This Important Bird Area doesn't get

Wilson's Snipe

the attention it deserves as a year-round birding location. Look for ducks like Common Goldeneye and Long-tailed Duck from the beach in winter; spring migrants include a number of warblers and other songbirds like Cedar Waxwing, American Redstart; thrushes and Carolina Wrens abound. There are three hiking trails, which cover 1.75 miles, and an observation blind next to a tidal lake for photographing waders. Like Caumsett, there are ticks and poison ivy and sometimes mosquitoes in summer. Located in the village of Lloyd Harbor off Target Rock Entrance Road. Vehicle entrance fee. Restrooms at the parking lot.

12 Target Rock Road, Huntington, NY 11743; 631-286-0485—headquarters office at Wertheim National Wildlife Refuge.

http://www.fws.gov/northeast/longislandrefuges

Tung Ting Pond and Mill Pond

If you are in the area during the winter, swing by Tung Ting Pond. This tiny spring-fed body of water, which derives its name from the Chinese restaurant that once operated here, remains open in colder weather, and can be particularly productive, showcasing some really nice waterfowl like Canvasbacks,

Redheads, Eurasian Wigeon, and Northern Pintail. For the best viewing, drive to the Chalet Motor Inn parking lot at 23 Centershore Road, in Centerport. After checking out Tung Ting Pond, be sure to walk across Centershore Road and take a look at the water flowing into Mill Pond, as the waters there could have some nice ducks as well. Mill Pond can also be viewed from a different angle at the dam on Mill Dam Road.

Sunken Meadow State Park

This extremely popular beach destination in Kings Park, with stunning views over Long Island Sound, is another Important Bird Area and a great four-season spot to search for shorebirds, songbirds, and winter waterfowl. Over a million visitors a year come here, so birders may prefer to plan a trip for the off-season to enjoy Sunken Meadow's three miles of beaches, glacier-sculpted cliffs, and twelve hundred acres of tidal wetlands and woods.

VIEWING SPOTS

Upon entering the park, as you cross the bridge over Sunken Meadow Creek, on the right you'll glimpse the part of the creek that is accessible to birders. You may be curious about the vast tidal flat on the left, western, side, but it is closed to visitors except via a *lookout near Parking Lot 2*. If that lot is closed, the best access to this area is to park in Lot 1 near the bridge and hike over. The lookout is easily found near an informational sign on the southern side of the road to the golf course.

Otherwise, park in the *eastern side of Lot 3* and bird the area there: beach, tidal marsh, grassy areas, trees. Make your way south, and you can now choose between crossing the bridge by the parking lot and birding along *the creek* from the southern side, or staying on the parking lot side and making your way along the bank. Each route has its advantages, and you can easily do both by making a loop. The birds seem to prefer the mudflats on the northern side but are often hard to see until you cross to the other side of the creek. In both cases you'll be scanning the thin strip of vegetation (laden with fruit in fall) bordering the creek, as well as the grassy fields, water, and exposed mud. Use the occasional fishermen's path through the underbrush for better views of the creek, but be prepared as poison ivy is abundant, and there are plenty of ticks.

A stroll along the *boardwalk* and a visit to the *woods* might also yield some interesting birds, depending on the season.

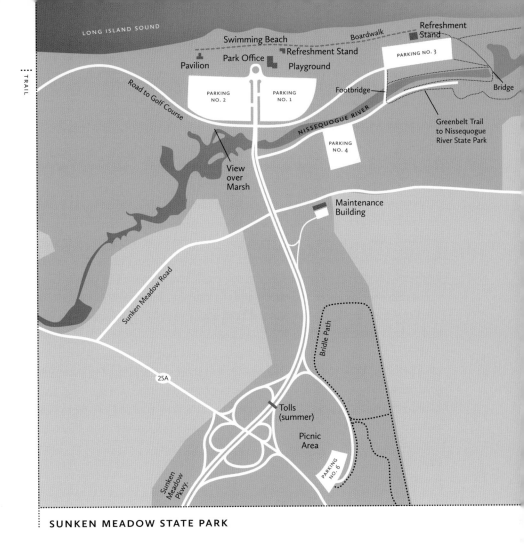

SUNKEN MEADOW STATE PARK

In spring, include a stop on the *southern side of Lot 4* and scan the willows for songbirds, especially warblers.

To maximize your visit during shorebird migration, plan to bird during low tide on the Nissequogue River.

A trip here is easily combined with visits to other parks in the area, such as *Nissequogue River State Park*. This location formerly was home to a psychiatric center, now fallen into disrepair, and it has amazing views of the waterway and offers canoe access. Nissequogue is home to colonial nesting herons and egrets in summer and a spot to visit during winter for ducks. Reached via Kings Park Boulevard.

You may also want to consider a stop any time of year at *Blydenburgh County Park*, which is described just below in this chapter.

KEY SPECIES BY SEASON

Spring Some shorebirds, but fall is better. In recent years, Blue-winged Teal have been making appearances. Good for songbirds, especially warblers. Wilson's Snipe may occasionally be seen.

Summer Nesting areas have been set aside for Piping Plovers and Least Terns. The woods and some of the other areas have a nice suite of breeding birds that include both cuckoos, Hairy Woodpecker, Willow Flycatcher, Blue-gray Gnatcatcher, Wood Thrush, White-eyed Vireo, Prairie, and other warblers, plus Rose-breasted Grosbeaks; Bank Swallows nest in the cliffs.

Fall Shorebirds begin appearing mid- to late July. Some raptors, especially Merlin; occasionally, Green-winged Teal; a nice spot for warblers, sparrows, and Indigo Bunting.

Winter Canada Geese and Mallards dominate but are mixed in with the usual array of Long-tailed Ducks, scaup, and scoters. Cackling, Pink-footed, or Greater White-fronted Geese are sometimes reported; Sanderlings and Dunlins. Overwintering songbirds can include Yellow-rumped Warblers, Fox and American Tree Sparrows, Golden-crowned Kinglets, and Purple Finches.

HABITAT

Beach, woods, tidal.

BEST TIME TO GO

Fall, winter, spring, summer—in that order.

HOW TO GET TO SUNKEN MEADOW STATE PARK

There is no practical access via public transportation. The nearest train station is the Long Island Rail Road stop at Kings Park. From there, it is a mile to the park entrance.

By Car Set your GPS for Route 25A and Sunken Meadow Parkway, Kings Park, NY 11754. From the Long Island Expressway or Northern Parkway, take the Sunken Meadow Parkway north to its end at the park entrance.

INFORMATION

Route 25A and Sunken Meadow Parkway, Kings Park, NY 11754

631-269-4333 / 631-269-5351

http://nysparks.com/parks/attachments/SunkenMeadowBirdChecklist
.pdf, where you can download a bird checklist

To get information on tides visit http://tides.mobilegeographics.com and search for Nissequogue River Entrance.

There is an entry fee in summer. Ample restroom facilities dot the park—the ones at Lot 1 are reliably open year-round. Concession stands offer food and drink in summer.

In the off-season, not all of the five parking lots may be open. Parking Lot 1 is nearest the park office, where you can pick up a map and consult a tide table if you haven't already done so online.

The bridge at the eastern side of Lot 3 used to be a dam but was destroyed by Hurricane Sandy. You may see references to it in descriptions of the park written before 2012.

OTHER THAN BIRDING

Nearly everything. Besides all the usual beach park activities of swimming, picnicking, fishing, basketball, softball, soccer, and golf (twenty-seven-hole course), there is a canoe and kayak launch at Lot 3, where one can explore not only Long Island Sound but the Nissequogue River as well. There are also cross-country races, horseback riding (BYO), cross-country skiing, model airplane flying, and a butterfly garden and playgrounds for the kids. Fees and permits may be required.

Overall this is a wonderful place to walk around in the off-season and take in the views of Long Island Sound, the bluffs, and the beach. Sunken Meadow Park is at the northern terminus of one of the Long Island Greenbelt Trails—gorgeous, especially in fall. Begin this hike at Lot 3.

An events calendar can be found at http://nysparks.com

Scope You'll be happy that you brought it.

Photos Easy.

Blydenburgh Park

This appealing six-hundred-acre wooded area in Smithtown and Hauppauge surrounds the headwaters of the Nissequogue River. Its main feature, Stump Pond, is one of the largest freshwater lakes on Long Island, and it's a great spot year-round—winter waterfowl, during migration, and during nesting season.

VIEWING SPOTS

We prefer to enter on the northern side, from New Mill Road, park in the lot (where restrooms are available), and hike the short trail to Stump Pond. It's worth a visit in late fall and winter to see Ring-necked Duck, Hooded Merganser, and Northern Pintail. Also check out the pond during migration when who knows what might turn up on the water. A good place in spring and summer for Wood Ducks and Red-winged Blackbirds. Look for warblers during spring and fall migration, and there is wonderful habitat for nesting woodland birds, too. This park hooks up with the Long Island Greenbelt Trail (see Sunken Meadow State Park in this chapter) and in fact is home to its headquarters. Enter at the end of New Mill Road in Smithtown. Park office: 631-854-3713.

David Weld Sanctuary

Enjoy the beauty and tranquillity of this 125-acre Nature Conservancy preserve in Nissequogue. Walk the three miles of trails along beautiful bluffs overlooking Smithtown Bay and through forests and wildflowers in spring and summer.

At the bluffs in summer, you can see the burrows made by Bank Swallows and watch them tend their nests. Check the forests during spring and fall migration for warblers, including Blue-winged; Eastern Towhee, and other songbirds. Great Horned Owl have been seen here. Summer brings some waders and shorebirds, and in winter, waterfowl in the bay include Red-breasted Merganser, Common Loon, and rafts of Brant and American Black Ducks.

Enter this North Shore preserve off Boney Lane (near the Horse Race Lane intersection). Parking is very limited.

Connetquot River State Park

Comprising nearly thirty-five hundred acres, this preserve in Oakdale is one of the largest wildlife havens on Long Island. The park was the site of a nineteenth-century sportsmen's club with its own trout hatchery, and fishing has long been a focus. But this preserve, which protects part of the ecologically unique Central Pine Barrens, also offers a wonderful habitat for birds. In fact, a section of this Important Bird Area is set aside so that birds can breed in peace. Wild trails and bridle paths meander through this attractive wooded park past ponds and along the Connetquot River, which flows through it.

VIEWING LOCATIONS

At the entrance, ask the ranger for a color-coded trail map. Birders will want to take the *Red Trail*, which leads from the parking lot and administrative buildings up the eastern side of *Main Pond* and meets the *Yellow Trail* at the hatchery. From there you can walk back to the parking lot along the pond's western side, making your way around the water and through the most productive birding areas. Main Pond is a haven for waterfowl and waders, but at some of the smaller, less visited ponds you may see Wood Ducks. Bluebirds can be seen in the *fields* around the pond. Note that many of the paths up to the Main Pond are often closed to anyone but fishermen holding permits.

For those in need of a longer experience with nature, take the *Blue Trail*, an 8.5-mile path through this unique ecosystem of sandy soil and somewhat stunted trees. The piney wooded areas in the eastern portion reached via the Blue Trail and Corwood Road are good spots for migrating songbirds, chickadees, and sometimes Screech Owls. With fifty miles of trails, there are plenty of opportunities to return to see more of the park. An erstwhile Northern Bobwhite release program means you'll have a very slight chance of spotting this once native breeder, but the attractions here are the freshwater and pine barrens communities of birds.

With plenty of standing water, Connetquot is an especially buggy place, partially compensated for by an abundance of dragonflies and damselflies in the summer. If you're concerned about ticks—and this is Long Island, after all—you could always walk up Hatchery Road and avoid the leaf litter and vegetation that ticks like.

Spring Migrating songbirds including over a dozen species of warblers; Eastern Bluebird, Eastern Towhees arrive.

Summer Wood Ducks; Osprey; wading birds; a really nice variety of breeding songbirds, including Yellow-billed Cuckoo, Eastern Wood-Pewee; Acadian, Alder, and Willow Flycatchers; White-eyed Vireo, Ovenbird; Field Sparrow, Scarlet Tanager, Baltimore Oriole, Yellow and Pine Warblers, Eastern Bluebird.

Fall Migrating songbirds, including warblers and sparrows; Purple Finch; ducks arrive.

Winter Overwintering waterfowl and a chance to see Ring-necked Ducks, Redheads, Canvasback, scaup, Gadwall, American Wigeon, Common and Hooded Mergansers, and more; plus the usual overwintering songbirds like Carolina Wren, White-throated Sparrow, Dark-eyed Junco, American Goldfinch, Tufted Titmouse; occasionally Screech Owl, Great Horned Owl.

HABITAT

Upland forest, freshwater lake, sandy pine barrens, wetlands.

BEST TIME TO GO

Summer, spring, fall, winter—in that order.

INFORMATION

4090 Sunrise Highway, Oakdale, NY

631-581-1005

http://nysparks.com/parks/attachments/ConnetquotRiverBirdChecklist
.pdf, where you can get a bird checklist

http://www.nynjtc.org

http://www.friendsofconnetquot.org

http://www.dec.ny.gov/outdoor

Please note that if you are driving from the west (that is, from New York City), you cannot make a left-hand turn from the Sunrise Highway into the preserve. You must overshoot the entrance and double back, exiting on your right from the east after Exit 47. Please also note that the address is sometimes listed as 3525 Sunrise Highway, but that will not take you to the park entrance. There is a per-car fee to enter.

The park is closed Monday and Tuesday, but even during those days the trails can be accessed by adjacent properties. Check the website for current opening days and hours.

OTHER THAN BIRDING

Fishing, bridle paths, cross-country skiing, and snowshoeing

Scope Would be helpful at the water, but not really necessary.

Photos Because the trees are stunted, the birds are not as high in the canopy as they are in other northeastern forests, but you must still contend with a certain amount of foliage, depending on the season. Photos are easy on the pond.

Bayard Cutting Arboretum State Park

Once the private home and property of a prominent New Yorker, this lovely arboretum located between Connetquot and Heckscher State Parks in Great River comprises over six hundred acres of trees and plantings, including large areas of native plants, ponds, and streamlets—all set aside exclusively for passive enjoyment so you won't have to battle bikers and horseback riders. Stroll to your heart's content on the many pathways around this park—it's difficult to make a wrong turn. In winter, take the Woodland Garden Walk past some ponds for waterfowl. Wander through the birdy native woodlands and the paths along the shore of the Connetquot River.

This park doesn't get a lot of attention, but it's tranquil and can be a good spot to see a variety of birds year-round. In addition to winter waterfowl, spring has some warblers and songbirds. In summer you may find songbirds like Baltimore Oriole, plus Osprey as well as interesting shorebirds like Solitary or Spotted Sandpiper in late summer. Fall has some migrating songbirds on their way through again, and a variety of sparrows are found here.

Along with birding you can visit the original home and relax in the café. The entire visit will be very civilized. There is a vehicle parking fee.

440 Montauk Highway, Great River, NY 11739; 631-581-1002
http://www.bayardcuttingarboretum.com

Heckscher State Park

Once the private property of two families, these fifteen hundred acres in East Islip on the South Shore were donated for a public recreational park and later designated part of an Important Bird Area. As a birding destination, it's not bad, but it is a popular beach playground, and during the summer season the park plays host to a million visitors per year. With twenty miles of hiking trails, and many other activities, there is a lot going on here in summer.

VIEWING SPOTS

As a birder, you may find this park's layout a bit unconventional. Some of the locals prefer to drive the loop, checking the parking lots for gulls. The grassy areas may have some sandpipers or other shorebirds. Park near the beach and walk to the water to see what may be passing through. The edges and wooded areas harbor a variety of songbirds, including vireos, flycatchers, and kingbird in spring. Interesting shorebirds including Least and Pectoral Sandpipers and Stilt Sandpiper are reported, but in summer you will have to contend with beachgoers in order to see them, so early morning visits are recommended. In fall, look for migrating raptors; Horned Lark in fall and early winter. During winter check for waterfowl, including uncommon geese. Greater White-fronted, Snow, and Ross's have been reported.

HOW TO GET TO HECKSCHER STATE PARK

Located south of the Sunrise Highway on Heckscher State Parkway.

INFORMATION

Heckscher Parkway Field 1, East Islip, NY 11730
631-581-2100
http://nysparks.com/parks/attachments/HeckscherBirdChecklist.pdf provides a checklist of birds

There are several bathrooms and large parking lots.
Heckscher has a lot of deer, which often bring ticks, so dress accordingly.

OTHER THAN BIRDING

Swimming, picnicking, camping, hiking, boating, playgrounds, playing fields, and other recreational activities.

Birders visit this unofficial collection of freshwater ponds, saltwater marshes, grasslands, sod farms, and a pine barrens—a unique sandy, scrubby, coniferous area—"for something different." The spots in this section are fairly close together and are often combined into a single day trip, depending on the season. The most productive are noted by an asterisk.

Wading River Marsh Preserve*

This attractive marshland in the hamlet of Wading River borders Long Island Sound. Follow Sound Road though a residential area facing the marsh, and stop as you see interesting birds and where there are safe pull-overs. Take Sound Road as far as it will go and park in the lot overlooking the beach. In winter, check for waterfowl including scoters and Long-tailed Ducks (which can be found in rafts by the hundreds) in the Sound; late summer hosts some migrating terns and shorebirds. Make a right out of the parking lot onto Creek Road, which will take you to the end of the peninsula bordering the Sound and where you will get sweeping views of this glorious salt marsh and the LILCO power plant. Several Osprey platforms support these nesting birds, and there is a Purple Martin house that is occasionally occupied.

http://www.nature.org

Wildwood State Park

If you have some extra time after Wading River Marsh Preserve in the winter, take a quick drive over to the beaches at this park, where you may find the likes of scoters and Long-tailed Ducks waiting for you in the Sound.

790 Hulse Landing Road, Wading River, NY 11792; 631-929-4314
http://nysparks.com

Hulse Landing Road*

If sparrows and raptors are your goal, give this road in Wading River a try from fall to early spring. Of special interest is the area between Sound Ave-

N

LONG ISLAND SOUND

FLANDERS BAY

27

Speonk–Riverhead Rd.

Sunrise Hwy.

Riverhead–Moriches Rd.

51

51

111

27

Peconic Bay Blvd.

Manor Lane

Tuthills

Church

West Ave.

Union

25

Main Rd.

Hubbard

RIVERHEAD

Cross River Dr.

Northville Tpke.

East Main

West Main

Riverleigh Ave.

Sound Ave.

Golden Triangle Sod Farms

Doctors Path

Reeves Ave.

Roanoke Ave.

Pulaski

Buffalo Farm

Horton Ave.

Old Country Rd.

Middle Country Rd.

PECONIC RIVER

Nugent Dr.

Osborne Ave.

Sound Ave.

Youngs Ave.

Middle Rd.

Riley

Twomey Ave.

Edwards

Fresh Pond Ave.

River Rd.

Mill Rd.

Long Island State Pine Barrens Preserve

East Manor

Jones Rd.

495

Wildwood State Park

Hulse Landing Rd.

Fresh Pond Ave.

Burman Blvd.

EPCAL

McKAY POND

SWAN LAKE

Grumman Blvd.

Calverton Preserve

Old River Rd.

Ponds Preserve Rd.

PRESTON'S POND

Schultz

Wading River Rd.

Wading River Marsh Preserve

N. Wading River Rd.

N. Country Rd.

Sound Ave.

Parker

Middle Country Rd.

25

Wading River Manor

Long Island Expy.

★ KEY SITE

CENTRAL SUFFOLK GRASSLANDS

Killdeer

nue and Parker Road, inland from the North Shore. Just south of Sunwood Drive to the east of Hulse Landing there is a dirt road with no name next to a public utility passage. Walk between the two sets of utility poles and scan the short trees, shrubs, and grass for sparrows. There are often overwintering White-crowned, American Tree, and Song Sparrows. Also check the taller trees for Rough-legged Hawk.

EPCAL*

EPCAL stands for Enterprise Park at Calverton, also known as Calverton Grasslands, formerly a site leased by the Grumman Corporation. Visit this abandoned airport to see raptors and grassland birds. Kestrels, Harrier, and Red-shouldered Hawk are just some of the birds of prey seen here year-round. The abandoned runway areas are a haven for grassland types such as Eastern Meadowlark, nesting Blue Grosbeaks, Prairie Warbler, Grasshopper and Field Sparrows, and occasionally nesting Bluebirds. Just inside the northern entrance to EPCAL off Middle Country Road there are trees where Short-eared Owls, if they are around, may be seen. At the southern entrance off Grumman Boulevard, *McKay Lake* may have some interesting waterfowl in winter and migrating shorebirds such as Killdeer. If you take this entrance, you will drive through the industrial center via Burman Boulevard to reach the airfield. There are barricades and No Trespassing signs posted on some of the runaways, but you can easily see the birds you want with a scope. Be aware that hunting Wild Turkey is permitted in season, so plan your visit to the airfield portion accordingly.

Calverton Ponds Preserve, Preston's Pond, and Swan Pond*

Just south of EPCAL is a series of several pretty ponds in different preserves. Access Preston's Pond off Grumman Boulevard. Just east of Wading River Manor Road, pull into the lot at the sign indicating a Fishing Access Site and take the trail through the pine barrens to the lake. While this area is not frequently birded, it might be worth a stop to see what is in there and take a stroll by the water. A little farther east on Grumman Boulevard is Swan Pond, which you can access by turning right on River Road. After about one-half mile, there is a pull-over from which you can take a short walk to the pond. Not a real hot spot, but it is on the way to Calverton Ponds Preserve, which you will not want to miss.

This gem of a Nature Conservancy property, Calverton Ponds Preserve, is also known as the Denis and Catherine Krusos Research Area and consists of three tranquil freshwater ponds—Block, Sandy, and Fox. All are accessed by beautiful hiking paths through the Central Pine Barrens forest. In early spring there may be quite a few Ring-necked Ducks. But summer is its best season, when shy Wood Duck are often seen. Yellow-billed Cuckoo, Eastern Wood-Pewee, Red-eyed Vireo, Yellow and Pine Warbler, Ovenbird, and other songbirds also nest here. The preserve is located west of Line Road on Old River Road; look for the Nature Conservancy sign for the entrance. Limited parking.

"The Buffalo Farm"

Located on Reeves and Roanoke Avenue in the town of Riverhead, this farm can be a stop on your way to the Golden Triangle (see below). In winter, you may find some interesting geese and overwintering sparrows. Often White-crowned Sparrows and Horned Lark are seen hanging out with the bison.

Golden Triangle Sod Farms*

There is no sign indicating the Golden Triangle. It's simply a local name for a drive-around of a large area of sod farms where you can look for American Golden-Plover, Upland and Pectoral Sandpipers, and other field-loving shorebirds during fall migration. Winter brings geese, and you never know what you may find here—Cackling Geese have been seen. Starting at the intersection of Northville Turnpike and Doctors Path in Riverhead, go north along Northville, left on Sound Avenue, and left on Doctors Path to complete the triangle. Some think it's best to cut off the first tip by going north on Northville, left on Cross River Drive (County Road 105), left on Sound Avenue, and left on Doctors, which makes it the Golden Trapezoid. Whichever geometric shape you decide to drive, look for safe pull-overs along the way.

THE SOUTH FORK AND SHELTER ISLAND

Shinnecock Bay and Inlet

Created during a storm in 1938, Shinnecock is one of five inlets that connect the Atlantic Ocean with bays behind the barrier islands protecting Suffolk County's South Shore. It is kept open by dredging, creating a number of often ephemeral sandbars that are popular with waterfowl. In winter be sure to stop here—a sure bet.

VIEWING SPOTS

Park at the end of Dune Road in Hampton Bays and make your way to the beach. From here there is a nice view of the ocean, and if there are any pelagics in spring and fall, this offers a good view. Check the ocean for whatever is stopping by. Search the channel and the Atlantic for waterfowl in winter. The waters around the jetty are legendary for the abundance of birds that congregate here from November through mid-March. There can be rarities such as King Eider, so take a good look. Also check out what is perched on the jetty tower and what is resting on the beach. Uncommon gulls, such as Glaucous and the odd Iceland Gull, are sometimes found loafing around here as well. Purple Sandpipers are sometimes seen on the rocks. Look for Harbor Seals hauled out in the bay.

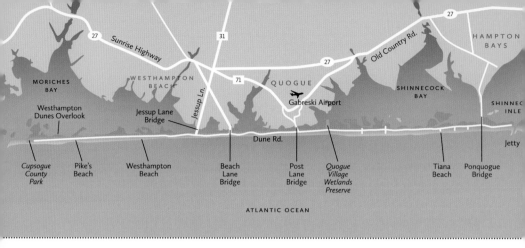

SHINNECOCK AND DUNE ROAD

KEY SPECIES BY SEASON

Spring Scoters persist, as do some waterfowl like Common Loon. Red-winged Blackbirds arrive.

Summer Occasional pelagics are sighted in the inlet. Waders, Osprey, a variety of shorebirds, including Piping Plover, Ruddy Turnstone, and Least Sandpiper. American Oystercatchers remain through fall. Black and Common Terns; Barn and Tree Swallows.

Fall Roseate Terns. Interesting sparrows arrive.

Winter Common Eider can number over a thousand, and look for all three scoters, which can number in the hundreds; Long-tailed Duck, Red-breasted Merganser, Red-throated Loon, Horned Grebe; occasional Razorbill; a chance at King Eider. Gulls possibly include Bonaparte's, Iceland or Glaucous; Purple Sandpiper. Some raptors, including Northern Harrier and Snowy Owl; Peregrine year-round.

HABITAT

Beach and marine.

BEST TIME TO GO

Winter and summer are the most productive, in that order.

HOW TO GET TO SHINNECOCK BAY AND INLET

When Dune Road dead-ends in the east, you have arrived.

If you want to reach the eastern side of the inlet, you will have to do a long drive-around. Much of the area is private, and it is not nearly as productive as the western side of the inlet.

Scope Bring it.

Photos You will need a long lens.

Dune Road

Dune Road is a popular fall and winter birding route on the South Fork of Long Island, although it can be good in any season. Its eastern terminus is Shinnecock Inlet, and the route runs the entire length of the barrier island ending at Cupsogue Beach County Park. There are numerous pull-overs and interesting bird areas along this nearly ten-mile stretch, which has bird-attracting mudflats and salt marshes on the bay side to the north, while the south side is beach and ocean. Dune Road is punctuated at each end with inlets, which have distinct personalities. As a result, throughout the drive you may see a wide variety of shorebirds, raptors, and migrants, depending on time of year and where you decide to stop.

Ruddy Turnstone

SUFFOLK COUNTY

Peregrine Falcon flying

VIEWING SPOTS

Begin at *Shinnecock Inlet*—see our description in this chapter.

From Shinnecock, drive west, making stops at available pull-overs. *Ponquogue Bridge* has a parking lot at the base of the bridge, with an overlook onto the canal. Check here in winter for waterfowl like Common Loon, Bufflehead, Red-breasted Merganser, and Long-tailed Ducks. Check out the marshy area under the bridge as well. Sparrows like the bushes along Dune Road, and you'll want to carefully check out that Song Sparrow, which might turn out to be an Ipswich Savannah.

Next stop, *Tiana Beach*, for more winter waterfowl. If you happen to be here in spring and summer, and it's possible to avoid the crowds, shorebirds such as Willet, Black-bellied Plover, and Ruddy Turnstone may be poking around.

Continue on to the *Quogue Village Wetlands Preserve*. It has parking for only a couple of cars, but it's worth a stop in spring if you can get in. Visit the boardwalk over the marsh and scan the ditches alongside for Clapper Rails. The stand of pines may net a few songbirds. And in summer, the beaches may have shorebirds. The Quogue Village Wetlands Preserve is not to be confused with the Quogue Wildlife Preserve on Old County Road. The Quogue Village

Wetlands Preserve is in the village of Quogue, and the tiny parking lot is across from 195 Dune Road.

At *Westhampton Dunes Overlook County Park* your best shot at seeing birds will probably be in late spring into September, when you may see the likes of Ruddy Turnstone, Red Knot, Least and White-rumped Sandpipers; Royal, Common, and Forster's Terns, and a variety of other shorebirds. But during the beach season you are going to be wrangling with others for parking spaces at this tiny lot, located west of the huge Pike's Beach lot on Dune Road near Cove Lane. Once parked in the lot, walk through the scrubby area to a viewing platform, where there are good views of Moriches Bay and ephemeral sandbars. It might not be as exciting to view birds on the mudflats from here, but it sure is a lot safer than Cupsogue. This parking lot is free. Bring the scope.

You can end your Dune Road expedition at *Cupsogue*—see our description in this chapter.

KEY SPECIES BY SEASON

Fall Migration stopover for shorebirds including Black-bellied Plover, Red Knot, Least and White-rumped Sandpipers; Common, Forster's, and Royal Terns.

Winter Significant overwintering location for waterfowl including Greater and Lesser Scaup, American Black Duck, Red-breasted Merganser, Brant, Common Goldeneye, Common Loon, Red-throated Loon, Mallard, Canada Goose, Long-tailed Duck, Canvasback, Bufflehead, Horned Grebe, Common Eider; Mute Swans, Black Scoter, Glaucous Gull, Great Cormorant, Great Blue Heron, Sanderlings, American Bittern, Clapper Rail, Peregrine Falcon, Cooper's Hawk, Northern Harrier, Boat-tailed Grackle, Savannah Sparrow (Ipswich subspecies), and American Tree and Song Sparrows.

Spring Migration stopover for shorebirds like dowitchers and Dunlins; terns arrive, with possible views of Black and Roseates.

Summer Osprey; shorebirds like Willet; terns that arrived in spring; Ruddy Turnstone into fall.

INFORMATION

Dune Road: http://www.libirding.com

Parking is not permitted on Dune Road, and your car may be towed if it is not parked in a designated beach pull-over or lot. From Labor Day to Memorial Day there is less traffic and easier parking. During summer, it is recommended that you get a beach sticker from the town of Southampton for parking at the town beaches.

Hunting is permitted in parts of this area, so plan your visit accordingly.

There are no public restrooms other than an occasional portable toilet along this route.

Restaurants are open during the season, and some are open year-round on weekends.

Scope Would be helpful.

Photos Birds may be at a distance.

Cupsogue Beach County Park

A visit to this popular barrier island summer beach at the end of Dune Road in Westhampton can be a worthwhile experience during much of the year, although it is known for its productive mudflats May through September. Located on the eastern side of Moriches Inlet, this Important Bird Area has shorebirds and terns, seabirds and waders spring through fall; and because it lures intrepid and knowledgeable birders, rarities are often reported. However, we hesitate to recommend it because a summer trip out to the mudflats must be made with EXTREME CAUTION, as it involves wading through moving tidewaters and walking on submerged sand and muck. You can easily sink into the gooey mud to which the birds are so attracted, and extrication will be difficult—especially without getting your equipment wet. If you are determined to make this journey, do so deliberatively, and make sure you go with someone who knows the "terrain."

VIEWING SPOTS

The key to a full and safe experience here in summer is to go at low tide and visit the mudflats first. Set out about one hour before low tide and come back to the beach by one hour after low tide. Be sure to use the tide table for the bay side (not the ocean side), Moriches Bay Coast Guard Station (http://www. surf-forecast.com/breaks/Cupsogue/tides/latest) to determine when dead low tide is. As noted, the journey can be dangerous if you don't know where to walk. Since you'll be wading and possibly sinking in mud, we recommend open rubber sandals that strap on securely—no flip flops! People with tough feet go barefoot. Mud boots will not work, as the water may fill your boots, making them a liability.

Once you've driven to the parking lot, try to find a slot at the end farthest away from the toll booth. Begin hiking due west on the sand-pack road, walking past the RV parking area on your left. Look for a sandy footpath on your right, which leads through the brush down to the *bayside beach* on the north side of the island. There may be a few birds in the scrub, but don't linger and miss the tide, as this will be your only opportunity to see the birds on the mudflats. At the beach, make a right along the sand. From here, let your knowledgeable companion show you where to walk and wade through the shallow water to the *mudflats* in front of you. Here's where the EXTREME CAUTION comes in. The way is slippery and may be unstable in areas, and what looks like sand could just be a thin layer brought in by the tide, and covering muck. Given all this, if you make it to the mudflats, and the birds happen to be there, you'll have a wonderful shorebird viewing experience.

Ephemeral sandbars crop up from time to time to the west and are popular with gulls and other birds. Check these out at low tide.

After you're back on dry land, and if you have the time, you can walk west to check out the beach birds at *the inlet*. This is your destination if you're at Cupsogue in winter, spring, or fall. When you return to the parking lot, you might want to walk to *the ocean-side beach* and scan the skies for migrating pelagics and seabirds like scoters.

If you're not up for such an expedition, visit the *Westhampton Dunes Overlook County Park*, on Dune Road between Cove Lane and Pike's Beach, where you can see ephemeral mudflats and sandbars from dry land. You'll see fewer birds, but it's safer. (More information in this chapter under Dune Road.)

KEY SPECIES BY SEASON

Spring Black-Bellied Plover can be seen year-round; Red Knot; occasional migrating scoters.

Summer Expect to see more common shorebirds and also Glossy Ibis, American Oystercatcher, Piping Plover, Marbled Godwit, Short-billed Dowitcher, Pectoral Sandpiper, and possible White-rumped Sandpiper; occasional Arctic, Black, and Roseate Terns; Clapper Rail; uncommonly, Wilson's Storm-Petrel and shearwaters.

Fall Migrating shorebirds and possibly Royal Tern; migrating raptors like Northern Harrier; occasional scoters.

Winter Overwintering waterfowl including loons, Red-breasted Merganser, and Long-tailed Ducks.

Beach, marine.

Summer shorebirds and terns are the highlight here. Waterfowl and shorebirds can be seen in winter, and migrating seabirds and shorebirds in spring and fall.

From the village of Westhampton Beach go south on Jessup Lane and cross the bridge. Turn right onto Dune Road. Follow Dune Road to its western terminus in Westhampton Beach.

Dune Road, Westhampton
631-852-8111 (seasonal)
http://www.suffolkcountyny.gov/Departments/Parks/Parks
There is limited and paid parking.
There are restrooms and a restaurant at the parking area.

A fire destroyed the beach pavilion in September 2014, but you can still swim in summer under the watch of a lifeguard and enjoy the restaurant in its temporary home. Camping, fishing, and scuba diving are also available.

Scope Most people bring one.

Photos You'll need a long lens.

Mecox Bay

Come to this bay and beach year-round for migrating shorebirds in spring and fall, nesting shorebirds in summer, and waterfowl in winter. It is best approached from the east on Dune Road in Bridgehampton; park in the lot at W. Scott Cameron Beach where the road dead-ends and head for the mudflats. Spring through fall you will see waders like Great Blue Heron and Great Egret; scan for shorebirds like Black-bellied Plover, Short-billed Dowitcher, Red Knot, and if you're lucky, Black and Caspian Terns in migration.

American Golden-Plover have been seen here from time to time. Roped-off sections on the beaches in summer protect nesting Piping Plovers. Also look for Least Tern and Black Skimmer; Willow Flycatcher and occasional warblers skulk in the shrubby areas. Bring your scope for good views of Red-throated and Common Loons and Bufflehead in winter and spring. Be aware that the roads surrounding the bay are private, and gaining access during the summer season may be difficult.

Mashomack Preserve on Shelter Island

Covering one-third of Shelter Island, Mashomack is a beautiful and diverse natural area of nearly twenty-one hundred acres that encompasses transitioning farmland, salt marsh, and tidal creeks, freshwater marsh, woods, and over twelve miles of shoreline. It also has one of the densest populations of breeding Osprey on the East Coast—one of many reasons the Nature Conservancy purchased the land in the 1980s. The preserve's mission is "to maintain its full array of natural communities and species, and to restore to health those that have been disturbed by human action," so only passive recreation is allowed, and not all parts are open to the public. Several trails of varying lengths are mapped out, which let you experience the different habitats up close. Don't make this a quick stop, as you are going to want to have enough time to see as much as you possibly can. Tranquil, scenic, immersive, teeming with wildlife—this Important Bird Area earns its name "Jewel of the Peconic."

VIEWING SPOTS

If you don't want to hike the full ten miles of trails, head out on the Red Trail, continue east on the Yellow Trail, and circle back on the Yellow Trail through the *meadow*. Those with more energy will want to also take in the Green Trail, which brings you above the salt marsh area known as *Log Cabin Creek* and around *Sanctuary Pond*. If you want the full experience, the Blue Trail will take you past some swamps and ponds by the beach and has a nice overlook not too far in from the Green Trail.

While on Shelter Island, you might want to drive out to the beaches at *Ram Island County Park*. But be mindful that in the summer, you'll have to be considerate of the nesting shorebirds along the beach. Unlike Mashomack, this is a quick stop.

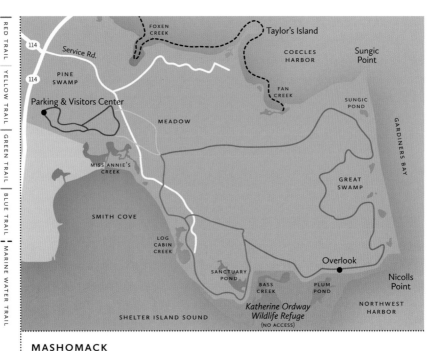

MASHOMACK

KEY SPECIES BY SEASON

Spring Waterfowl linger through April; some warblers, including Yellow-rumped; White-eyed Vireo, Eastern Towhee.

Summer Least Terns, Great Crested Flycatcher, White-eyed Vireo; Tree, Barn, and Bank Swallows; Eastern Bluebird, Field Sparrow and a thriving array of nesters, including Blue-winged and Prairie Warblers, Orchard and Baltimore Orioles, Osprey; Bald Eagles in an off-limits part of the preserve.

Fall Shorebirds; a great time for warblers; a nice set of sparrows, including the occasional White-crowned, plus Purple Finch.

Winter It's a long hike to see saltwater waterfowl and Sanderling. Along the way you'll find the usual overwintering songbirds. The resident Great Horned Owls are more easily seen.

HABITAT

Tidal creeks, mature oak woodlands, fields, freshwater marshes, and coastline.

BEST TIME TO GO

A truly excellent spot in spring, summer, and fall.

HOW TO GET TO MASHOMACK PRESERVE ON SHELTER ISLAND

Shelter Island is accessed only by ferry. One operates out of Greenport and is known as the North Ferry, since it comes from the North Fork of Long Island; the other is out of North Haven but is called the South Ferry, since it comes from the south. Both take cars and pedestrians, and both can get backed up in the summer.

INFORMATION

47 South Ferry Road, Shelter Island, NY 11964
631-749-1001

While this Nature Conservancy–operated preserve is open year-round, it is open only on the weekends in January. A full listing of times can be found at http://www.nature.org

The Harman Hawkins Visitor Center has exhibits and a small store and restrooms. Often a naturalist is on hand to answer questions. Nature programs and guided hikes are also offered.

The sleepy little coastal town of Shelter Island has great places to stay and dine and play on the beach if you want to make a weekend of it.

THE NORTH FORK

There are a number of parks and shore areas on the North Fork, and despite their often wild nature and natural beauty, as birding locations they are not generally considered hot spots. Winter is arguably the best time of year here, and you can expect to find loons, scoters, Long-tailed Ducks, and most of the usual waterfowl. Summer brings Osprey and nesting shorebirds; fall, woodpeckers and raptors; spring, some migrants, but they tend to wind up on the South Fork. Individual seasons aren't really the point here—it is the sum of the seasons that makes it interesting for birders.

Below is a list of parks, beaches, and wild areas on the North Fork that are known birding spots. They are strung along New York State Highway 25, also called Main Road; we list them from east to west. The most productive are marked with an asterisk. There is an active community of preservationists on the North Fork working toward land conservation. Hopefully we can look forward to additional bird-friendly locations opening up.

Orient Point County Park*

A sort of mini-Montauk, this Important Bird Area occupies the tip of the tine on the North Fork. Most people visit in winter, parking at the end of State Highway 25 next to the ferry dock. Assuming that you want to use the trails to make a loop, start with a hike along the lovely pebble beach to look for sea ducks. At land's end you'll get nice views of Long Island Sound, the merging waters of the Sound and the Atlantic Ocean (in an upwelling of water known as the Plum Gut), the Orient Light, and Plum Island (see our description following). The trek back to the car takes you through brushy woods on a dirt road where you can find a variety of songbirds. In summer you'll find terns, Osprey, swallows. Because of its scenic nature, this is a good, short hike to make with non-birders.

Plum Island

From Orient Point you may be tempted to try to figure out a way to get to this island, as it looks like it should be prime birding. In fact, it is a designated Important Bird Area with critical habitat of beach, wetlands, grasslands, and forest, and it protects nesting birds by denying entry to all except for U.S. Department of Agriculture researchers. However, tours are sometimes offered through bird clubs and local environmental organizations. Give it a try if you're curious, but remember that you'll ride around the whole time on a bus and won't be allowed to walk around and explore. Breeding Osprey, Bank Swallows, Piping Plovers, Roseate and Common Terns benefit, along with a number of overwintering waterfowl and other native animals and plants. http://www.preserveplumisland.org

Orient Beach State Park*

While it is a separate preserve from Orient Point County Park, the two are lumped together as an Important Bird Area, and often both are visited in the same day. Orient Beach has its own entrance on State Highway 25 on

Common Eider

the south side, at 40000 Main Road, Orient, New York, about one-quarter mile before the entrance to Orient Point County Park when coming from the west. Consisting of a long strip of barrier island, Orient Beach has a varied and precious habitat of maritime forest, freshwater and saltwater marshes, and beach. A car path of about two and a half miles brings you to a large parking lot about halfway down the island. From here, you can hike a nature trail through the woods and along the beach to Long Beach Point. You'll find colonial breeding birds, including Piping Plovers and terns; Osprey; and in winter, waterfowl. You will pay an entry fee in summer and share your experience with beachgoers, kayakers, and windsurfers. Restrooms are open year-round.

Ruth Oliva Preserve at Dam Pond

Sometimes simply known as Dam Pond, this thirty-six-acre preserve combines grassland, forest, and tidal estuary, so the variety of birds found here ranges widely—songbirds, waders, raptors, ducks, shorebirds. It has a reputation for being better than most of the North Fork for early spring migrants, warblers, and fall sparrows and finches; expect Osprey and some nesting

SUFFOLK COUNTY

255

warblers in summer. The entrance is on Cove Beach Road, off Main Road (State Highway 25) in East Marion. A trail map is available at http://www.southoldtownny.gov

Inlet Pond County Park

The North Fork Audubon Society stewards this wooded preserve with beachfront access to Long Island Sound. There are one and a half miles of trails. The entrance is at 65275 Route 48, Greenport.

Moore's Woods*

This hardwood forest is a hidden gem of spring, summer, and fall songbird habitat. Look for warblers, Great Crested Flycatchers, Red-eyed Vireos, Scarlet Tanagers, and thrushes—some of which nest here as well. You'll find nuthatches, woodpeckers, and more common birds year-round. Moore's Woods is also known for its rare native orchids and other botanical delights. The entrance is from Moore's Lane between Main Road and Route 48, Greenport.

Arshamomaque Preserve

This Greenport preserve has a variety of habitats, including meadows, bogs, forest, and freshwater ponds. There are Nesting Wood Ducks, Great Horned Owl, and Osprey; Woodcock make appearances in spring; summer sees flycatchers; and you will find forest birds year-round. This is also a good place to see butterflies and muskrats. Access the trailhead on Chapel Lane north of Main Road.

Arshamomaque Pond Preserve*

Part of the Long Island Pine Barrens Maritime Reserve, Arshamomaque Pond Preserve has trails that take you to a saltwater tidal pond, open meadows, and through forest and bogs. The pond hosts overwintering waterfowl. Summer has waders, Osprey, terns, some flycatchers, Yellow Warblers, and Baltimore Orioles. In fall, a variety of migrating shorebirds pass through. Year-round, see woodpeckers and forest dwelling-birds. Access the trailhead on the north side of Main Road just east of Port of Egypt. The pond is also sometimes referred to as Hashamomuck.

Cedar Beach County Park

This park is located in Southold and consists of wetlands, mudflats, and open beach. Plan to go at low tide to visit the exposed mudflats. In winter, search for Long-tailed Ducks and Common Goldeneye among other waterfowl; in summer, Piping Plover, Osprey, gulls, swallows. A variety of migrating shorebirds, including Black-bellied Plovers, Whimbrel, and Ruddy Turnstone, pass through here—especially late summer into fall. Waterfowl and shorebirds account for the majority of activity. The park is accessed via Main Bayview Road, which becomes Cedar Beach Road.

Goldsmith's Inlet Park*

Known for its great views of Long Island Sound, Goldsmith's is a small park with mature woodlands, tidal wetlands, and beach. Check out both sides of the inlet and pond. The trail that leads to the beach also goes through the wooded area and past salt marshes and is accessed off Soundview. Take Soundview west and make a right on Mill Lane to reach the west side of the pond and inlet. In winter, good numbers of scoters and Long-tailed Ducks can be seen, along with Common Loon, mergansers, and Horned Grebe; summer brings nesting Osprey, Common Tern, Belted Kingfisher, nesting Prairie Warbler. Piping Plover may breed on these busy beaches; some songbirds. The trailhead is on Soundview Avenue one-quarter mile east of Mill Lane in Southold.

Nassau Point of Little Hog Neck

This spit of beach in Cutchogue juts into the Peconic Bay where Nassau Point Road dead-ends. Winter may net some waterfowl; in summer, Osprey and shorebirds.

Downs Farm Preserve

These fifty-one acres of coastal and woodland habitat are operated by the town of Southold and located at 23800 Main Road (Highway 25) in the village of Cutchogue. Nesting Yellow Warbler and Eastern Towhee; woodpeckers, some waders, and relatively common waterfowl.

Marratooka Lake Park (Marratooka Pond)

Bring your scope to this freshwater lake in Mattituck, which might have interesting winter waterfowl. Access is via New Suffolk Avenue, south of Main Road in Mattituck.

Husing Pond Preserve

A freshwater pond in a small preserve makes Husing ideal for a quick stop. Look for winter ducks, including Ring-necked, and Bufflehead in Horton Creek. In spring, some songbirds, including Warbling Vireo; in summer, Osprey, Yellow Warbler, swallows. There is a nice assortment of sparrows throughout the year; some warblers pass through on migration both directions. The preserve is located off Peconic Bay Boulevard in Laurel.

Laurel Lake

This freshwater lake surrounded by woodlands and an open grassland in Southold hosts waterfowl and migrating and nesting birds. Spring through fall sees a variety of warblers; summer has nesting songbirds and Osprey. Check the treetops in the meadow behind the parking lot during migration. The lake is on the north side of Main Road in Laurel.

MONTAUK PENINSULA

Long Island's Montauk Peninsula is the easternmost stretch of land in New York State, jutting out into the Atlantic and ending at Montauk Point, which is a must-see spot prized by birders. While you're nearby, stop by East Hampton's Hook Pond for winter waterfowl (this is also convenient to a trip to Shinnecock). There are other places in the area to look for birds on the peninsula, but they are worth going to only if you have plenty of time on your hands after visiting Montauk Point.

Montauk Point State Park and Camp Hero State Park

It may be just over one hundred miles from New York City, but this easternmost part of New York State seems a world away. A legendary fisherman's

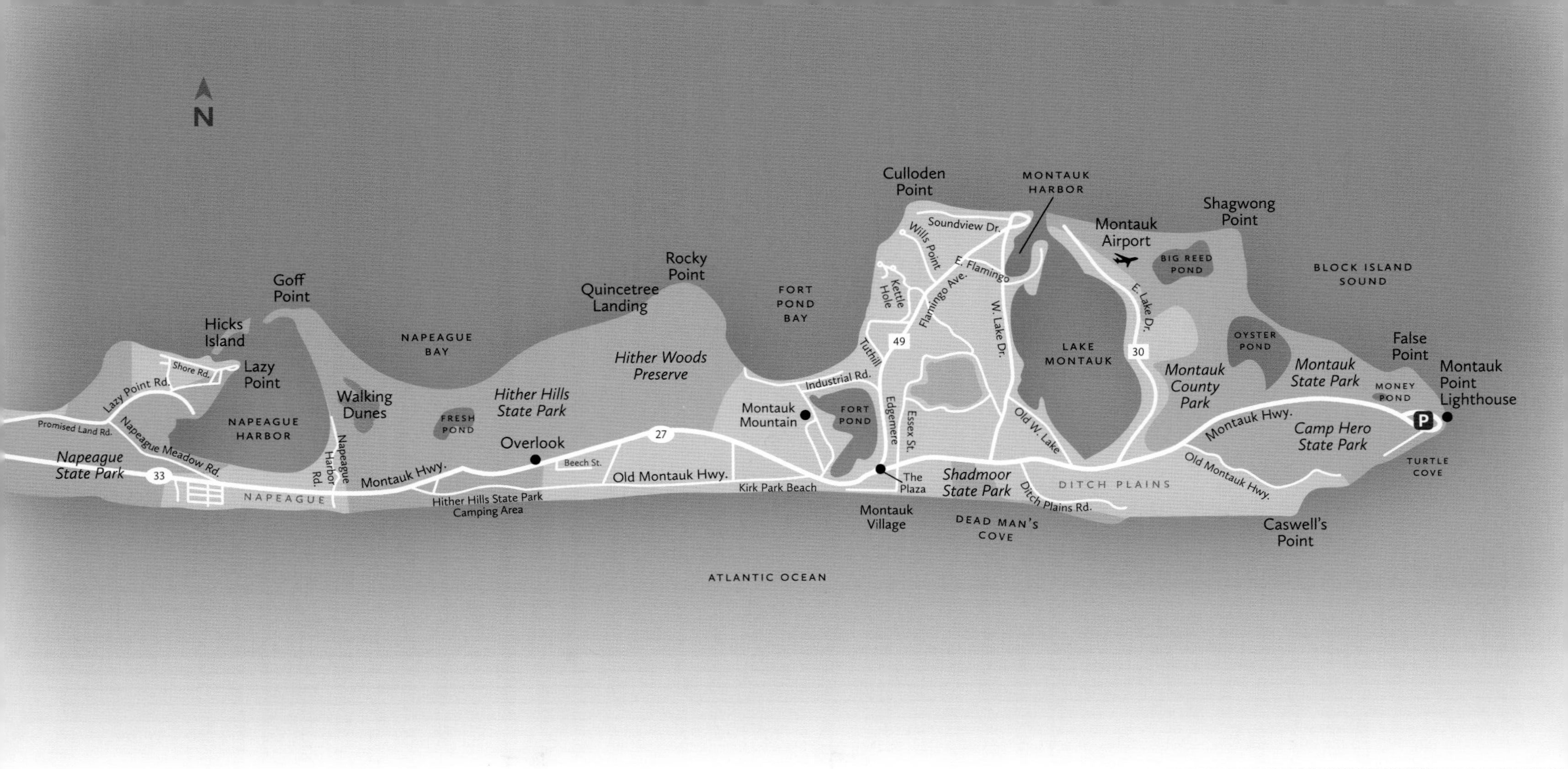

MONTAUK PENINSULA

Hicks Island
Goff Point
Lazy Point
Rocky Point
Quincetree Landing
Culloden Point
Shagwong Point
False Point
Montauk Point Lighthouse
Caswell's Point

BLOCK ISLAND SOUND

Montauk State Park
Camp Hero State Park
TURTLE COVE
MONEY POND

Montauk Airport
E. Lake Dr.
30
Montauk County Park
OYSTER POND
BIG REED POND
Montauk Hwy.
Old Montauk Hwy.
DITCH PLAINS

MONTAUK HARBOR
Soundview Dr.
Wills Point
E. Flamingo Ave.
W. Lake Dr.
Old W. Lake
LAKE MONTAUK
Flamingo Ave.
Kettle Hole
49
Tuthill
Essex St.
Edgemere
The Plaza
Shadmoor State Park
DEAD MAN'S COVE
Ditch Plains Rd.
Montauk Village

FORT POND BAY
Industrial Rd.
FORT POND
Montauk Mountain
Kirk Park Beach

ATLANTIC OCEAN

Hither Woods Preserve
27
Old Montauk Hwy.
Overlook
Beech St.
Hither Hills State Park
FRESH POND

NAPEAGUE BAY
Walking Dunes
Napeague Harbor Rd.
Montauk Hwy.
Hither Hills State Park Camping Area

NAPEAGUE HARBOR
NAPEAGUE
Shore Rd.
Lazy Point Rd.
Promised Land Rd.
Napeague Meadow Rd.
33
Napeague State Park

N

paradise, it holds a similar allure for birders seeking waterfowl. Remote and rocky, Montauk Point, with its rolling hills, lakes, boulder-strewn shores, and picturesque scenery, juts far into the Atlantic. It is a must-do winter birding expedition, as the species range and sheer quantity of waterfowl seen here are unsurpassed anywhere on Long Island.

Camp Hero State Park is just west of Montauk Point and comprises 415 acres of beach and bluffs, woodland, and freshwater wetlands. This former military base is now a registered National Historic Site and has trails that take you through the woods, over creeks, and past decommissioned bunkers. You'll get spectacular views of the water and further opportunities to see waterfowl, although the birding is better at Montauk Point.

These two parks are part of a seventy-eight-hundred-acre Important Bird Area on the eastern tip of Long Island benefiting a wide variety of species, including waterfowl, sea ducks, pelagics, and even breeding songbirds. If you are willing to brave the crowds of nature lovers and the abundance of ticks in warmer months, you will see a few migrating and nesting songbirds and some interesting shorebirds that pass through, like Roseate Terns.

VIEWING SPOTS

Once you have parked in the huge lot at *Montauk State Park*, make your way toward the lighthouse and down to the beach. Low tide allows more room for your scope, and more gulls will be present, but it is not strictly necessary to be there at low tide. The richest haul will probably be by the *lighthouse* itself, where the Atlantic meets Block Island Sound and the converging waters bring fish to the surface. The beach is built of boulders here, so wear appropriate shoes, and know that it can be mighty windy. Travel westward along the beach to check out the Harbor Seals that haul out just offshore. The surrounding *woods* may harbor migrants during fall.

Montauk puts you near the birds, but if you want to make another stop, check out what's doing at *Camp Hero*. This Atlantic-facing park can have plenty of waterfowl as well, but you see them from atop a cliff, and they are much farther away. A scope is even more of a necessity here.

Lake Montauk may also have some interesting waterfowl, especially near the mouth of the lake on the north side. Here's also your chance to look for Purple Sandpipers and Dunlin at the jetties.

On your way from Montauk, visit *Napeague and Lazy Point*. Make a right after Hither Hills State Park on Napeague Meadow Road (County Road 33)

Shorebirds in flight during storm

for a drive-and-stop through the marsh. At the T-intersection with Lazy Point Road, make a right, and drive to Lazy Point itself for views on Long Island Sound and the chance to see more waterfowl and maybe an uncommon gull.

KEY SPECIES BY SEASON

Winter A waterfowl bonanza, including the chance to see King Eider; Common Eider may be seen by the thousands; Harlequin Duck; Surf, White-winged, and Black Scoters—sometimes in very large numbers; Common and Red-throated Loons; Red-necked Grebe, Razorbill, Black-legged Kittiwake; there can be large numbers of Long-tailed Ducks and Northern Gannets. Sometimes an unusual gull like Bonaparte's, Iceland, or Glaucous will be found.

Spring Warblers and songbirds; Common and Roseate Terns spring and into fall, sometimes in large numbers; Red-eyed Vireo pass through; Baltimore Oriole; Eastern Towhee, plus other sparrows are here year-round.

Summer Osprey; a chance of seeing migrating pelagics like gannets, petrels, and shearwaters beginning in July and through the fall. Shearwaters, if they appear, may be in large flocks.

Fall Waterfowl arrive. Great Cormorants arrive for overwintering; a few shorebirds are seen; Caspian Terns have made sporadic appearances over the

SUFFOLK COUNTY

261

years; woodpeckers including Northern Flicker; Tree Swallows; sparrows, including Dark-eyed Junco.

HABITAT

Beach, rocky shore, open water, woods.

BEST TIME TO GO

Winter is best, but Montauk is also recommended in the fall. The birding is good in other seasons, but you'll have to deal with ticks and crowds to find birds that you might more easily see somewhere else.

HOW TO GET TO MONTAUK STATE PARK

Montauk State Park is at the eastern end of the Montauk Highway (State Highway 27). For Camp Hero, leave the parking lot and continue on the one-way loop until you can turn left as though returning to the parking lot, and make a right onto the road to Camp Hero. For Lake Montauk, return to Highway 27 and make a right on East Lake Drive (County Road 30), which will end at the mouth of the lake.

INFORMATION

http://nysparks.com
Montauk State Park
2000 Montauk Highway, Montauk, NY 11954
631-668-3781

Camp Hero State Park
1898 Montauk Highway, Montauk, NY 11954
631-668-3781

Parking fees are not collected in the off-season at Montauk Point State Park, and the restrooms are open; the concession stand and historic lighthouse are closed. No facilities at Camp Hero or at Lake Montauk.

In some older guidebooks you will see reference to the Montauk Dump. It's been closed for many years and is now part of Hither Hills State Park.

The hunting seasons run from roughly December 1 to February 28.

Be aware that in winter Montauk can be extremely cold and windy, so dress accordingly.

Scope Yes—this is a necessity.

Song Sparrow

Photos Easiest at Montauk Point; at Camp Hero, the waterfowl are much farther away.

Shadmoor State Park and Ditch Plains

Shadmoor State Park is a ninety-nine-acre expanse of low shadbush, bluffs, freshwater wetlands, and birdwatching platforms. Two World War II concrete bunkers are a reminder of the role these cliffs played in protecting the coastline. It might be an interesting next stop, and even though it's close to Hither Hills State Park, Shadmoor is a completely different experience.

VIEWING SPOTS

Stop at the informational plaque when you get there for a full appreciation of this special geological formation. Don't miss the elevated platforms for good views over the water. Bank Swallows nest in the cliffs, and some birders feel it's worth a visit in summer for this alone. In winter, you can stand on the cliffs with your scope and look for waterfowl, although just east of here is the mother lode of winter waterfowl at Montauk.

Be warned that there is no shade during the entire walk, and there are *two* World War II bunkers, which can be confusing, because they are virtually identical. When you come upon the other one on your return to the parking lot, continue along the trail and make the first right to get back to the car.

Shadmoor is adjacent to the surfing destination *Ditch Plains*. Squeeze into the lot at Ditch Plains and take a short stroll along the beach to view the cliffs and birdlife. Gaining access to the top of the cliffs is easy from the beach. This is also a good option for those who don't feel like schlepping their scopes through the preserve.

900 Montauk Highway, Montauk, NY 11954; 631-668-3781
http://nysparks.com, where you can download a trail map

Hither Hills State Park

Hither Hills is a three-thousand-acre park with forests, a freshwater pond, and high cliffs giving stunning views of Napeague Bay. It also happens to be something of a warbler magnet in spring and fall. Summer is its best season, with lots of nesting birds. Located on the eastern part of Long Island near Montauk, the park runs the full width of the peninsula from Napeaque Bay to the Atlantic Ocean, although for birds, you're going to visit only the part north of Montauk Highway.

VIEWING SPOTS

Most people visit the southern portion of Hither Hills in the summer for the beaches and playgrounds, or the Walking Dunes on the western side. If it's birds you are after, your best bet will be to head for the woods and freshwater lake. Reputedly, this is the only place where you can see all eight breeding species of warbler on Long Island, including Blue-winged, Prairie, and Oven-bird, plus Baltimore Oriole and Red-eyed Vireo. In other seasons, you can find songbird migrants in spring and fall. Since hunting is permitted in winter, we don't recommend Hither Hills for birdwatching during that season.

HOW TO GET TO HITHER HILLS STATE PARK

The Overlook Parking Lot, which is where birders flock, is on Montauk Highway (Highway 27) on the north side, just west of Beech Street. The entrance to the campground and beach is off Old Montauk Highway and not accessible from Highway 27.

Montauk, NY 11954 (The official park address of 164 Old Montauk Highway will take you to the campground and beach, not the birding area described here)

631-668-2554

http://nysparks.com (provides a bird conservation area map)

There are restrooms and picnic facilities at Hither Hills in the campground area.

Be aware that the hunting season runs roughly from November 1 to February 28, which you should take into consideration if you're thinking of combining your visit with a winter outing to Montauk Point.

Be prepared for ticks in season.

OTHER THAN BIRDING

Fishing, year-round with permit; hunting; bridle paths; swimming and camping in the southern portion of the park seasonally; biking, cross-country skiing. Fees apply to many of these activities, but you'll pay nothing if you park in the Overlook lot or at Napeague Harbor and hike or birdwatch.

The birding route does not include Hither Hills's most interesting scenic feature, the Walking Dunes, located on the western side of the park. The dunes are actually quite beautiful and worth a look. Kids will love the chance to roll down the slopes. But rather than trying to incorporate the dunes into your bird walk, think about parking at the trailhead off Napeague Harbor Road and visiting them in a separate mini-hike.

Hook Pond

This is a quaint sixty-four-acre pond near the beach in East Hampton that provides refuge for a number of interesting ducks, geese, and swans in the winter. Park in the beach parking lot at the end of Ocean Avenue and double back a few yards, making a right onto the sand-pack road. The pond is best in winter, especially at the end of the day when waterfowl that have been out foraging return to roost. Be mindful that the pond is surrounded by private property, including the Maidstone Golf Club.

Scope Recommended.

Photos A long lens would be helpful, as waterfowl can be far away.

8 Species Accounts

The following bar charts indicate the relative abundance of birds that occur annually across the region. Lists of Rarities and Accidental Species follow. Data are taken primarily from the Cornell Lab of Ornithology's eBird.org citizen science database, *The Second Atlas of Breeding Birds in New York State* by Kevin J. McGowan (Ithaca, NY: Comstock Publishing Associates, 2008), the AOU *Checklist of North American Birds*, 7th Checklist, 54th Supplement, September 2013, and from the Annual Reports of New York State Avian Records Committee. The birds are listed in taxonomic order according to the AOU *Checklist*.

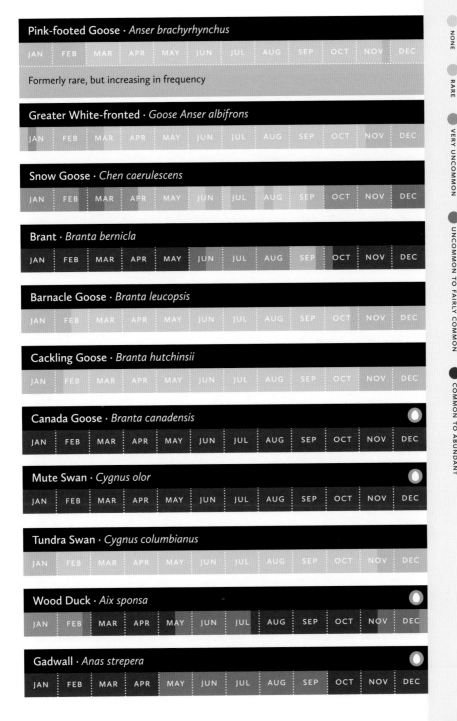

Pink-footed Goose · *Anser brachyrhynchus*

| JAN | FEB | MAR | APR | MAY | JUN | JUL | AUG | SEP | OCT | NOV | DEC |

Formerly rare, but increasing in frequency

Greater White-fronted · *Goose Anser albifrons*

| JAN | FEB | MAR | APR | MAY | JUN | JUL | AUG | SEP | OCT | NOV | DEC |

Snow Goose · *Chen caerulescens*

| JAN | FEB | MAR | APR | MAY | JUN | JUL | AUG | SEP | OCT | NOV | DEC |

Brant · *Branta bernicla*

| JAN | FEB | MAR | APR | MAY | JUN | JUL | AUG | SEP | OCT | NOV | DEC |

Barnacle Goose · *Branta leucopsis*

| JAN | FEB | MAR | APR | MAY | JUN | JUL | AUG | SEP | OCT | NOV | DEC |

Cackling Goose · *Branta hutchinsii*

| JAN | FEB | MAR | APR | MAY | JUN | JUL | AUG | SEP | OCT | NOV | DEC |

Canada Goose · *Branta canadensis*

| JAN | FEB | MAR | APR | MAY | JUN | JUL | AUG | SEP | OCT | NOV | DEC |

Mute Swan · *Cygnus olor*

| JAN | FEB | MAR | APR | MAY | JUN | JUL | AUG | SEP | OCT | NOV | DEC |

Tundra Swan · *Cygnus columbianus*

| JAN | FEB | MAR | APR | MAY | JUN | JUL | AUG | SEP | OCT | NOV | DEC |

Wood Duck · *Aix sponsa*

| JAN | FEB | MAR | APR | MAY | JUN | JUL | AUG | SEP | OCT | NOV | DEC |

Gadwall · *Anas strepera*

| JAN | FEB | MAR | APR | MAY | JUN | JUL | AUG | SEP | OCT | NOV | DEC |

NONE
RARE
VERY UNCOMMON
UNCOMMON TO FAIRLY COMMON
COMMON TO ABUNDANT

BREEDS IN NEW YORK CITY AND ON LONG ISLAND

Eurasian Wigeon · *Anas penelope*

JAN	FEB	MAR	APR	MAY	JUN	JUL	AUG	SEP	OCT	NOV	DEC

American Wigeon · *Anas americana*

JAN	FEB	MAR	APR	MAY	JUN	JUL	AUG	SEP	OCT	NOV	DEC

American Black Duck · *Anas rubripes*

JAN	FEB	MAR	APR	MAY	JUN	JUL	AUG	SEP	OCT	NOV	DEC

Mallard · *Anas platyrhynchos*

JAN	FEB	MAR	APR	MAY	JUN	JUL	AUG	SEP	OCT	NOV	DEC

Blue-winged Teal · *Anas discors*

JAN	FEB	MAR	APR	MAY	JUN	JUL	AUG	SEP	OCT	NOV	DEC

Northern Shoveler · *Anas clypeata*

JAN	FEB	MAR	APR	MAY	JUN	JUL	AUG	SEP	OCT	NOV	DEC

Northern Pintail · *Anas acuta*

JAN	FEB	MAR	APR	MAY	JUN	JUL	AUG	SEP	OCT	NOV	DEC

Green-winged Teal · *Anas crecca*

JAN	FEB	MAR	APR	MAY	JUN	JUL	AUG	SEP	OCT	NOV	DEC

Canvasback · *Aythya valisineria*

JAN	FEB	MAR	APR	MAY	JUN	JUL	AUG	SEP	OCT	NOV	DEC

Redhead · *Aythya americana*

JAN	FEB	MAR	APR	MAY	JUN	JUL	AUG	SEP	OCT	NOV	DEC

Ring-necked Duck · *Aythya collaris*

JAN	FEB	MAR	APR	MAY	JUN	JUL	AUG	SEP	OCT	NOV	DEC

Greater Scaup · *Aythya marila*

JAN	FEB	MAR	APR	MAY	JUN	JUL	AUG	SEP	OCT	NOV	DEC

Lesser Scaup · *Aythya affinis*

JAN	FEB	MAR	APR	MAY	JUN	JUL	AUG	SEP	OCT	NOV	DEC

King Eider · *Somateria spectabilis*

JAN	FEB	MAR	APR	MAY	JUN	JUL	AUG	SEP	OCT	NOV	DEC

Common Eider · *Somateria mollissima*

JAN	FEB	MAR	APR	MAY	JUN	JUL	AUG	SEP	OCT	NOV	DEC

Harlequin Duck · *Histrionicus histrionicus*

JAN	FEB	MAR	APR	MAY	JUN	JUL	AUG	SEP	OCT	NOV	DEC

Surf Scoter · *Melanitta perspicillata*

JAN	FEB	MAR	APR	MAY	JUN	JUL	AUG	SEP	OCT	NOV	DEC

White-winged Scoter · *Melanitta fusca*

JAN	FEB	MAR	APR	MAY	JUN	JUL	AUG	SEP	OCT	NOV	DEC

Black Scoter · *Melanitta americana*

JAN	FEB	MAR	APR	MAY	JUN	JUL	AUG	SEP	OCT	NOV	DEC

Long-tailed Duck · *Clangula hyemalis*

JAN	FEB	MAR	APR	MAY	JUN	JUL	AUG	SEP	OCT	NOV	DEC

Bufflehead · *Bucephala albeola*

JAN	FEB	MAR	APR	MAY	JUN	JUL	AUG	SEP	OCT	NOV	DEC

Common Goldeneye · *Bucephala clangula*

JAN	FEB	MAR	APR	MAY	JUN	JUL	AUG	SEP	OCT	NOV	DEC

Barrow's Goldeneye · *Bucephala islandica*

JAN	FEB	MAR	APR	MAY	JUN	JUL	AUG	SEP	OCT	NOV	DEC

Hooded Merganser · *Lophodytes cucullatus*

JAN	FEB	MAR	APR	MAY	JUN	JUL	AUG	SEP	OCT	NOV	DEC

BREEDS IN NEW YORK CITY AND ON LONG ISLAND

NONE

RARE

VERY UNCOMMON

UNCOMMON TO FAIRLY COMMON

COMMON TO ABUNDANT

Common Merganser · *Mergus merganser*

JAN	FEB	MAR	APR	MAY	JUN	JUL	AUG	SEP	OCT	NOV	DEC

Red-breasted Merganser · *Mergus serrator*

JAN	FEB	MAR	APR	MAY	JUN	JUL	AUG	SEP	OCT	NOV	DEC

Ruddy Duck · Oxyura *jamaicensis*

JAN	FEB	MAR	APR	MAY	JUN	JUL	AUG	SEP	OCT	NOV	DEC

Northern Bobwhite · *Colinus virginianus*

JAN	FEB	MAR	APR	MAY	JUN	JUL	AUG	SEP	OCT	NOV	DEC

Often a caged or reared bird in most parts of New York State, but there is an established breeding colony in Suffolk County

Ring-necked Pheasant · *Phasianus colchicu*

JAN	FEB	MAR	APR	MAY	JUN	JUL	AUG	SEP	OCT	NOV	DEC

Breeding population in serious decline. Most birds seen in the field are likely those reared and released for hunting.

Wild Turkey · *Meleagris gallopavo*

JAN	FEB	MAR	APR	MAY	JUN	JUL	AUG	SEP	OCT	NOV	DEC

Red-throated Loon · *Gavia stellata*

JAN	FEB	MAR	APR	MAY	JUN	JUL	AUG	SEP	OCT	NOV	DEC

Common Loon · *Gavia immer*

JAN	FEB	MAR	APR	MAY	JUN	JUL	AUG	SEP	OCT	NOV	DEC

Pied-billed Grebe · *Podilymbus podiceps*

JAN	FEB	MAR	APR	MAY	JUN	JUL	AUG	SEP	OCT	NOV	DEC

Horned Grebe · *Podiceps auritus*

JAN	FEB	MAR	APR	MAY	JUN	JUL	AUG	SEP	OCT	NOV	DEC

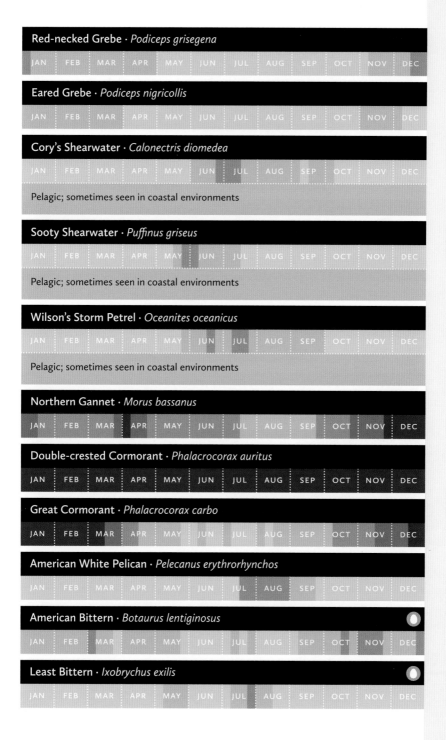

Red-necked Grebe · *Podiceps grisegena*

| JAN | FEB | MAR | APR | MAY | JUN | JUL | AUG | SEP | OCT | NOV | DEC |

Eared Grebe · *Podiceps nigricollis*

| JAN | FEB | MAR | APR | MAY | JUN | JUL | AUG | SEP | OCT | NOV | DEC |

Cory's Shearwater · *Calonectris diomedea*

| JAN | FEB | MAR | APR | MAY | JUN | JUL | AUG | SEP | OCT | NOV | DEC |

Pelagic; sometimes seen in coastal environments

Sooty Shearwater · *Puffinus griseus*

| JAN | FEB | MAR | APR | MAY | JUN | JUL | AUG | SEP | OCT | NOV | DEC |

Pelagic; sometimes seen in coastal environments

Wilson's Storm Petrel · *Oceanites oceanicus*

| JAN | FEB | MAR | APR | MAY | JUN | JUL | AUG | SEP | OCT | NOV | DEC |

Pelagic; sometimes seen in coastal environments

Northern Gannet · *Morus bassanus*

| JAN | FEB | MAR | APR | MAY | JUN | JUL | AUG | SEP | OCT | NOV | DEC |

Double-crested Cormorant · *Phalacrocorax auritus*

| JAN | FEB | MAR | APR | MAY | JUN | JUL | AUG | SEP | OCT | NOV | DEC |

Great Cormorant · *Phalacrocorax carbo*

| JAN | FEB | MAR | APR | MAY | JUN | JUL | AUG | SEP | OCT | NOV | DEC |

American White Pelican · *Pelecanus erythrorhynchos*

| JAN | FEB | MAR | APR | MAY | JUN | JUL | AUG | SEP | OCT | NOV | DEC |

American Bittern · *Botaurus lentiginosus*

| JAN | FEB | MAR | APR | MAY | JUN | JUL | AUG | SEP | OCT | NOV | DEC |

Least Bittern · *Ixobrychus exilis*

| JAN | FEB | MAR | APR | MAY | JUN | JUL | AUG | SEP | OCT | NOV | DEC |

BREEDS IN NEW YORK CITY AND ON LONG ISLAND

NONE

RARE

VERY UNCOMMON

UNCOMMON TO FAIRLY COMMON

COMMON TO ABUNDANT

Great Blue Heron · *Ardea herodias* ⬤

| JAN | FEB | MAR | APR | MAY | JUN | JUL | AUG | SEP | OCT | NOV | DEC |

Great Egret · *Ardea alba* ⬤

| JAN | FEB | MAR | APR | MAY | JUN | JUL | AUG | SEP | OCT | NOV | DEC |

Snowy Egret · *Egretta thula* ⬤

| JAN | FEB | MAR | APR | MAY | JUN | JUL | AUG | SEP | OCT | NOV | DEC |

Little Blue Heron · *Egretta caerulea* ⬤

| JAN | FEB | MAR | APR | MAY | JUN | JUL | AUG | SEP | OCT | NOV | DEC |

Tricolored Heron · *Egretta tricolor* ⬤

| JAN | FEB | MAR | APR | MAY | JUN | JUL | AUG | SEP | OCT | NOV | DEC |

Green Heron · *Butorides virescens* ⬤

| JAN | FEB | MAR | APR | MAY | JUN | JUL | AUG | SEP | OCT | NOV | DEC |

Black-crowned Night-Heron · *Nycticorax nycticorax* ⬤

| JAN | FEB | MAR | APR | MAY | JUN | JUL | AUG | SEP | OCT | NOV | DEC |

Yellow-crowned Night-Heron · *Nyctanassa violacea* ⬤

| JAN | FEB | MAR | APR | MAY | JUN | JUL | AUG | SEP | OCT | NOV | DEC |

Glossy Ibis · *Plegadis falcinellus* ⬤

| JAN | FEB | MAR | APR | MAY | JUN | JUL | AUG | SEP | OCT | NOV | DEC |

Black Vulture · *Coragyps atratus* ⬤

| JAN | FEB | MAR | APR | MAY | JUN | JUL | AUG | SEP | OCT | NOV | DEC |

Turkey Vulture · *Cathartes aura* ⬤

| JAN | FEB | MAR | APR | MAY | JUN | JUL | AUG | SEP | OCT | NOV | DEC |

Osprey · *Pandion haliaetus* ⬤

| JAN | FEB | MAR | APR | MAY | JUN | JUL | AUG | SEP | OCT | NOV | DEC |

Bald Eagle · *Haliaeetus leucocephalus*

JAN	FEB	MAR	APR	MAY	JUN	JUL	AUG	SEP	OCT	NOV	DEC

Northern Harrier · *Circus cyaneus*

JAN	FEB	MAR	APR	MAY	JUN	JUL	AUG	SEP	OCT	NOV	DEC

Sharp-shinned Hawk · *Accipiter striatus*

JAN	FEB	MAR	APR	MAY	JUN	JUL	AUG	SEP	OCT	NOV	DEC

Cooper's Hawk · *Accipiter cooperii*

JAN	FEB	MAR	APR	MAY	JUN	JUL	AUG	SEP	OCT	NOV	DEC

Red-shouldered Hawk · *Buteo lineatus*

JAN	FEB	MAR	APR	MAY	JUN	JUL	AUG	SEP	OCT	NOV	DEC

Broad-winged Hawk · *Buteo platypterus*

JAN	FEB	MAR	APR	MAY	JUN	JUL	AUG	SEP	OCT	NOV	DEC

Red-tailed Hawk · *Buteo jamaicensis*

JAN	FEB	MAR	APR	MAY	JUN	JUL	AUG	SEP	OCT	NOV	DEC

Rough-legged Hawk · *Buteo lagopus*

JAN	FEB	MAR	APR	MAY	JUN	JUL	AUG	SEP	OCT	NOV	DEC

Clapper Rail · *Rallus longirostris*

JAN	FEB	MAR	APR	MAY	JUN	JUL	AUG	SEP	OCT	NOV	DEC

Sora · *Porzana carolina*

JAN	FEB	MAR	APR	MAY	JUN	JUL	AUG	SEP	OCT	NOV	DEC

American Coot · *Fulica americana*

JAN	FEB	MAR	APR	MAY	JUN	JUL	AUG	SEP	OCT	NOV	DEC

American Avocet · *Recurvirostra americana*

JAN	FEB	MAR	APR	MAY	JUN	JUL	AUG	SEP	OCT	NOV	DEC

BREEDS IN NEW YORK CITY AND ON LONG ISLAND

NONE

RARE

VERY UNCOMMON

UNCOMMON TO FAIRLY COMMON

COMMON TO ABUNDANT

American Oystercatcher · *Haematopus palliatus*

| JAN | FEB | MAR | APR | MAY | JUN | JUL | AUG | SEP | OCT | NOV | DEC |

Black-bellied Plover · *Pluvialis squatarola*

| JAN | FEB | MAR | APR | MAY | JUN | JUL | AUG | SEP | OCT | NOV | DEC |

American Golden-Plover · *Pluvialis dominica*

| JAN | FEB | MAR | APR | MAY | JUN | JUL | AUG | SEP | OCT | NOV | DEC |

Semipalmated Plover · *Charadrius semipalmatus*

| JAN | FEB | MAR | APR | MAY | JUN | JUL | AUG | SEP | OCT | NOV | DEC |

Piping Plover · *Charadrius melodus*

| JAN | FEB | MAR | APR | MAY | JUN | JUL | AUG | SEP | OCT | NOV | DEC |

Killdeer · *Charadrius vociferus*

| JAN | FEB | MAR | APR | MAY | JUN | JUL | AUG | SEP | OCT | NOV | DEC |

Spotted Sandpiper · *Actitis macularius*

| JAN | FEB | MAR | APR | MAY | JUN | JUL | AUG | SEP | OCT | NOV | DEC |

Solitary Sandpiper · *Tringa solitaria*

| JAN | FEB | MAR | APR | MAY | JUN | JUL | AUG | SEP | OCT | NOV | DEC |

Greater Yellowlegs · *Tringa melanoleuca*

| JAN | FEB | MAR | APR | MAY | JUN | JUL | AUG | SEP | OCT | NOV | DEC |

Willet · *Tringa semipalmata*

| JAN | FEB | MAR | APR | MAY | JUN | JUL | AUG | SEP | OCT | NOV | DEC |

Lesser Yellowlegs · *Tringa flavipes*

| JAN | FEB | MAR | APR | MAY | JUN | JUL | AUG | SEP | OCT | NOV | DEC |

Whimbrel · *Numenius phaeopus*

| JAN | FEB | MAR | APR | MAY | JUN | JUL | AUG | SEP | OCT | NOV | DEC |

Hudsonian Godwit · *Limosa haemastica*

JAN	FEB	MAR	APR	MAY	JUN	JUL	AUG	SEP	OCT	NOV	DEC

Marbled Godwit · *Limosa fedoa*

JAN	FEB	MAR	APR	MAY	JUN	JUL	AUG	SEP	OCT	NOV	DEC

Ruddy Turnstone · *Arenaria interpres*

JAN	FEB	MAR	APR	MAY	JUN	JUL	AUG	SEP	OCT	NOV	DEC

Red Knot · *Calidris canutus*

JAN	FEB	MAR	APR	MAY	JUN	JUL	AUG	SEP	OCT	NOV	DEC

Stilt Sandpiper · *Calidris himantopus*

JAN	FEB	MAR	APR	MAY	JUN	JUL	AUG	SEP	OCT	NOV	DEC

Sanderling · *Calidris alba*

JAN	FEB	MAR	APR	MAY	JUN	JUL	AUG	SEP	OCT	NOV	DEC

Purple Sandpiper · *Calidris maritima*

JAN	FEB	MAR	APR	MAY	JUN	JUL	AUG	SEP	OCT	NOV	DEC

Baird's Sandpiper · *Calidris bairdii*

JAN	FEB	MAR	APR	MAY	JUN	JUL	AUG	SEP	OCT	NOV	DEC

Least Sandpiper · *Calidris minutilla*

JAN	FEB	MAR	APR	MAY	JUN	JUL	AUG	SEP	OCT	NOV	DEC

White-rumped Sandpiper · *Calidris fuscicollis*

JAN	FEB	MAR	APR	MAY	JUN	JUL	AUG	SEP	OCT	NOV	DEC

Buff-breasted Sandpiper · *Tryngites subruficollis*

JAN	FEB	MAR	APR	MAY	JUN	JUL	AUG	SEP	OCT	NOV	DEC

Pectoral Sandpiper · *Calidris melanotos*

JAN	FEB	MAR	APR	MAY	JUN	JUL	AUG	SEP	OCT	NOV	DEC

BREEDS IN NEW YORK CITY AND ON LONG ISLAND

- NONE
- RARE
- VERY UNCOMMON
- UNCOMMON TO FAIRLY COMMON
- COMMON TO ABUNDANT

Semipalmated Sandpiper · *Calidris pusilla*

JAN	FEB	MAR	APR	MAY	JUN	JUL	AUG	SEP	OCT	NOV	DEC

Western Sandpiper · *Calidris mauri*

JAN	FEB	MAR	APR	MAY	JUN	JUL	AUG	SEP	OCT	NOV	DEC

Short-billed Dowitcher · *Limnodromus griseus*

JAN	FEB	MAR	APR	MAY	JUN	JUL	AUG	SEP	OCT	NOV	DEC

Long-billed Dowitcher · *Limnodromus scolopaceus*

JAN	FEB	MAR	APR	MAY	JUN	JUL	AUG	SEP	OCT	NOV	DEC

Wilson's Snipe · *Gallinago delicata*

JAN	FEB	MAR	APR	MAY	JUN	JUL	AUG	SEP	OCT	NOV	DEC

Possible breeder

American Woodcock · *Scolopax minor*

JAN	FEB	MAR	APR	MAY	JUN	JUL	AUG	SEP	OCT	NOV	DEC

Wilson's Phalarope · *Phalaropus tricolor*

JAN	FEB	MAR	APR	MAY	JUN	JUL	AUG	SEP	OCT	NOV	DEC

Razorbill · *Alca torda*

JAN	FEB	MAR	APR	MAY	JUN	JUL	AUG	SEP	OCT	NOV	DEC

Bonaparte's Gull · *Chroicocephalus philadelphia*

JAN	FEB	MAR	APR	MAY	JUN	JUL	AUG	SEP	OCT	NOV	DEC

Laughing Gull · *Leucophaeus atricilla*

JAN	FEB	MAR	APR	MAY	JUN	JUL	AUG	SEP	OCT	NOV	DEC

Ring-billed Gull · *Larus delawarensis*

JAN	FEB	MAR	APR	MAY	JUN	JUL	AUG	SEP	OCT	NOV	DEC

Herring Gull · *Larus argentatus*

JAN	FEB	MAR	APR	MAY	JUN	JUL	AUG	SEP	OCT	NOV	DEC

Iceland Gull · *Larus glaucoides*

JAN	FEB	MAR	APR	MAY	JUN	JUL	AUG	SEP	OCT	NOV	DEC

Lesser Black-backed Gull · *Larus fuscus*

JAN	FEB	MAR	APR	MAY	JUN	JUL	AUG	SEP	OCT	NOV	DEC

Glaucous Gull · *Larus hyperboreus*

JAN	FEB	MAR	APR	MAY	JUN	JUL	AUG	SEP	OCT	NOV	DEC

Great Black-backed Gull · *Larus marinus*

JAN	FEB	MAR	APR	MAY	JUN	JUL	AUG	SEP	OCT	NOV	DEC

Least Tern · *Sternula antillarum*

JAN	FEB	MAR	APR	MAY	JUN	JUL	AUG	SEP	OCT	NOV	DEC

Gull-billed Tern · *Gelochelidon nilotica*

JAN	FEB	MAR	APR	MAY	JUN	JUL	AUG	SEP	OCT	NOV	DEC

Caspian Tern · *Hydroprogne caspia*

JAN	FEB	MAR	APR	MAY	JUN	JUL	AUG	SEP	OCT	NOV	DEC

Black Tern · *Chlidonias niger*

JAN	FEB	MAR	APR	MAY	JUN	JUL	AUG	SEP	OCT	NOV	DEC

Roseate Tern · *Sterna dougallii*

JAN	FEB	MAR	APR	MAY	JUN	JUL	AUG	SEP	OCT	NOV	DEC

Common Tern · *Sterna hirundo*

JAN	FEB	MAR	APR	MAY	JUN	JUL	AUG	SEP	OCT	NOV	DEC

Forster's Tern · *Sterna forsteri*

JAN	FEB	MAR	APR	MAY	JUN	JUL	AUG	SEP	OCT	NOV	DEC

Royal Tern · *Thalasseus maximus*

JAN	FEB	MAR	APR	MAY	JUN	JUL	AUG	SEP	OCT	NOV	DEC

BREEDS IN NEW YORK CITY AND ON LONG ISLAND

NONE
RARE
VERY UNCOMMON
UNCOMMON TO FAIRLY COMMON
COMMON TO ABUNDANT

Black Skimmer · *Rynchops niger*

JAN	FEB	MAR	APR	MAY	JUN	JUL	AUG	SEP	OCT	NOV	DEC

Rock Pigeon · *Columba livia*

JAN	FEB	MAR	APR	MAY	JUN	JUL	AUG	SEP	OCT	NOV	DEC

Mourning Dove · *Zenaida macroura*

JAN	FEB	MAR	APR	MAY	JUN	JUL	AUG	SEP	OCT	NOV	DEC

Yellow-billed Cuckoo · *Coccyzus americanus*

JAN	FEB	MAR	APR	MAY	JUN	JUL	AUG	SEP	OCT	NOV	DEC

Black-billed Cuckoo · *Coccyzus erythropthalmus*

JAN	FEB	MAR	APR	MAY	JUN	JUL	AUG	SEP	OCT	NOV	DEC

Eastern Screech Owl · *Megascops asio*

JAN	FEB	MAR	APR	MAY	JUN	JUL	AUG	SEP	OCT	NOV	DEC

Great Horned Owl · *Bubo virginianus*

JAN	FEB	MAR	APR	MAY	JUN	JUL	AUG	SEP	OCT	NOV	DEC

Snowy Owl · *Bubo scandiacus*

JAN	FEB	MAR	APR	MAY	JUN	JUL	AUG	SEP	OCT	NOV	DEC

Long-eared Owl · *Asio otus*

JAN	FEB	MAR	APR	MAY	JUN	JUL	AUG	SEP	OCT	NOV	DEC

Very rarely breeds here

Short-eared Owl · *Asio flammeus*

JAN	FEB	MAR	APR	MAY	JUN	JUL	AUG	SEP	OCT	NOV	DEC

Very rarely breeds here

Northern Saw-whet Owl · *Aegolius acadicus*

JAN	FEB	MAR	APR	MAY	JUN	JUL	AUG	SEP	OCT	NOV	DEC

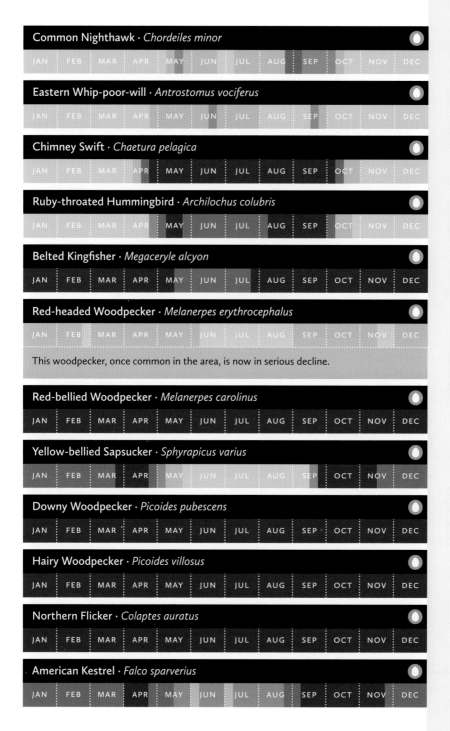

Common Nighthawk · *Chordeiles minor*

JAN · FEB · MAR · APR · MAY · JUN · JUL · AUG · SEP · OCT · NOV · DEC

Eastern Whip-poor-will · *Antrostomus vociferus*

JAN · FEB · MAR · APR · MAY · JUN · JUL · AUG · SEP · OCT · NOV · DEC

Chimney Swift · *Chaetura pelagica*

JAN · FEB · MAR · APR · MAY · JUN · JUL · AUG · SEP · OCT · NOV · DEC

Ruby-throated Hummingbird · *Archilochus colubris*

JAN · FEB · MAR · APR · MAY · JUN · JUL · AUG · SEP · OCT · NOV · DEC

Belted Kingfisher · *Megaceryle alcyon*

JAN · FEB · MAR · APR · MAY · JUN · JUL · AUG · SEP · OCT · NOV · DEC

Red-headed Woodpecker · *Melanerpes erythrocephalus*

JAN · FEB · MAR · APR · MAY · JUN · JUL · AUG · SEP · OCT · NOV · DEC

This woodpecker, once common in the area, is now in serious decline.

Red-bellied Woodpecker · *Melanerpes carolinus*

JAN · FEB · MAR · APR · MAY · JUN · JUL · AUG · SEP · OCT · NOV · DEC

Yellow-bellied Sapsucker · *Sphyrapicus varius*

JAN · FEB · MAR · APR · MAY · JUN · JUL · AUG · SEP · OCT · NOV · DEC

Downy Woodpecker · *Picoides pubescens*

JAN · FEB · MAR · APR · MAY · JUN · JUL · AUG · SEP · OCT · NOV · DEC

Hairy Woodpecker · *Picoides villosus*

JAN · FEB · MAR · APR · MAY · JUN · JUL · AUG · SEP · OCT · NOV · DEC

Northern Flicker · *Colaptes auratus*

JAN · FEB · MAR · APR · MAY · JUN · JUL · AUG · SEP · OCT · NOV · DEC

American Kestrel · *Falco sparverius*

JAN · FEB · MAR · APR · MAY · JUN · JUL · AUG · SEP · OCT · NOV · DEC

BREEDS IN NEW YORK CITY AND ON LONG ISLAND

NONE

RARE

VERY UNCOMMON

UNCOMMON TO FAIRLY COMMON

COMMON TO ABUNDANT

Merlin · *Falco columbarius*

JAN	FEB	MAR	APR	MAY	JUN	JUL	AUG	SEP	OCT	NOV	DEC

Peregrine Falcon · *Falco peregrinus*

JAN	FEB	MAR	APR	MAY	JUN	JUL	AUG	SEP	OCT	NOV	DEC

Monk Parakeet · *Myiopsitta monachus*

JAN	FEB	MAR	APR	MAY	JUN	JUL	AUG	SEP	OCT	NOV	DEC

Olive-sided Flycatcher · *Contopus cooperi*

JAN	FEB	MAR	APR	MAY	JUN	JUL	AUG	SEP	OCT	NOV	DEC

Eastern Wood-Pewee · *Contopus virens*

JAN	FEB	MAR	APR	MAY	JUN	JUL	AUG	SEP	OCT	NOV	DEC

Yellow-bellied Flycatcher · *Empidonax flaviventris*

JAN	FEB	MAR	APR	MAY	JUN	JUL	AUG	SEP	OCT	NOV	DEC

Acadian Flycatcher · *Empidonax virescens*

JAN	FEB	MAR	APR	MAY	JUN	JUL	AUG	SEP	OCT	NOV	DEC

Willow Flycatcher · *Empidonax traillii*

JAN	FEB	MAR	APR	MAY	JUN	JUL	AUG	SEP	OCT	NOV	DEC

Least Flycatcher · *Empidonax minimus*

JAN	FEB	MAR	APR	MAY	JUN	JUL	AUG	SEP	OCT	NOV	DEC

Eastern Phoebe · *Sayornis phoebe*

JAN	FEB	MAR	APR	MAY	JUN	JUL	AUG	SEP	OCT	NOV	DEC

Great Crested Flycatcher · *Myiarchus crinitus*

JAN	FEB	MAR	APR	MAY	JUN	JUL	AUG	SEP	OCT	NOV	DEC

Eastern Kingbird · *Tyrannus tyrannus*

JAN	FEB	MAR	APR	MAY	JUN	JUL	AUG	SEP	OCT	NOV	DEC

White-eyed Vireo · *Vireo griseus*

JAN	FEB	MAR	APR	MAY	JUN	JUL	AUG	SEP	OCT	NOV	DEC

Yellow-throated Vireo · *Vireo flavifrons*

JAN	FEB	MAR	APR	MAY	JUN	JUL	AUG	SEP	OCT	NOV	DEC

Blue-headed Vireo · *Vireo solitarius*

JAN	FEB	MAR	APR	MAY	JUN	JUL	AUG	SEP	OCT	NOV	DEC

Warbling Vireo · *Vireo gilvus*

JAN	FEB	MAR	APR	MAY	JUN	JUL	AUG	SEP	OCT	NOV	DEC

Philadelphia Vireo · *Vireo philadelphicus*

JAN	FEB	MAR	APR	MAY	JUN	JUL	AUG	SEP	OCT	NOV	DEC

Red-eyed Vireo · *Vireo olivaceus*

JAN	FEB	MAR	APR	MAY	JUN	JUL	AUG	SEP	OCT	NOV	DEC

Blue Jay · *Cyanocitta cristata*

JAN	FEB	MAR	APR	MAY	JUN	JUL	AUG	SEP	OCT	NOV	DEC

American Crow · *Corvus brachyrhynchos*

JAN	FEB	MAR	APR	MAY	JUN	JUL	AUG	SEP	OCT	NOV	DEC

Fish Crow · *Corvus ossifragus*

JAN	FEB	MAR	APR	MAY	JUN	JUL	AUG	SEP	OCT	NOV	DEC

Common Raven · *Corvus corax*

JAN	FEB	MAR	APR	MAY	JUN	JUL	AUG	SEP	OCT	NOV	DEC

Very rare breeder. Sightings of ravens are becoming increasingly common.

Horned Lark · *Eremophila alpestris*

JAN	FEB	MAR	APR	MAY	JUN	JUL	AUG	SEP	OCT	NOV	DEC

Northern Rough-winged Swallow · *Stelgidopteryx serripennis*

JAN	FEB	MAR	APR	MAY	JUN	JUL	AUG	SEP	OCT	NOV	DEC

BREEDS IN NEW YORK CITY AND ON LONG ISLAND

NONE

RARE

VERY UNCOMMON

UNCOMMON TO FAIRLY COMMON

COMMON TO ABUNDANT

Purple Martin · *Progne subis*

JAN	FEB	MAR	APR	MAY	JUN	JUL	AUG	SEP	OCT	NOV	DEC

Tree Swallow · *Tachycineta bicolor*

JAN	FEB	MAR	APR	MAY	JUN	JUL	AUG	SEP	OCT	NOV	DEC

Bank Swallow · *Riparia riparia*

JAN	FEB	MAR	APR	MAY	JUN	JUL	AUG	SEP	OCT	NOV	DEC

Barn Swallow · *Hirundo rustica*

JAN	FEB	MAR	APR	MAY	JUN	JUL	AUG	SEP	OCT	NOV	DEC

Black-capped Chickadee · *Poecile atricapillus*

JAN	FEB	MAR	APR	MAY	JUN	JUL	AUG	SEP	OCT	NOV	DEC

Tufted Titmouse · *Baeolophus bicolor*

JAN	FEB	MAR	APR	MAY	JUN	JUL	AUG	SEP	OCT	NOV	DEC

Red-breasted Nuthatch · *Sitta canadensis*

JAN	FEB	MAR	APR	MAY	JUN	JUL	AUG	SEP	OCT	NOV	DEC

White-breasted Nuthatch · *Sitta carolinensis*

JAN	FEB	MAR	APR	MAY	JUN	JUL	AUG	SEP	OCT	NOV	DEC

Brown Creeper · *Certhia americana*

JAN	FEB	MAR	APR	MAY	JUN	JUL	AUG	SEP	OCT	NOV	DEC

House Wren · *Troglodytes aedon*

JAN	FEB	MAR	APR	MAY	JUN	JUL	AUG	SEP	OCT	NOV	DEC

Winter Wren · *Troglodytes hiemalis*

JAN	FEB	MAR	APR	MAY	JUN	JUL	AUG	SEP	OCT	NOV	DEC

Marsh Wren · *Cistothorus palustris*

JAN	FEB	MAR	APR	MAY	JUN	JUL	AUG	SEP	OCT	NOV	DEC

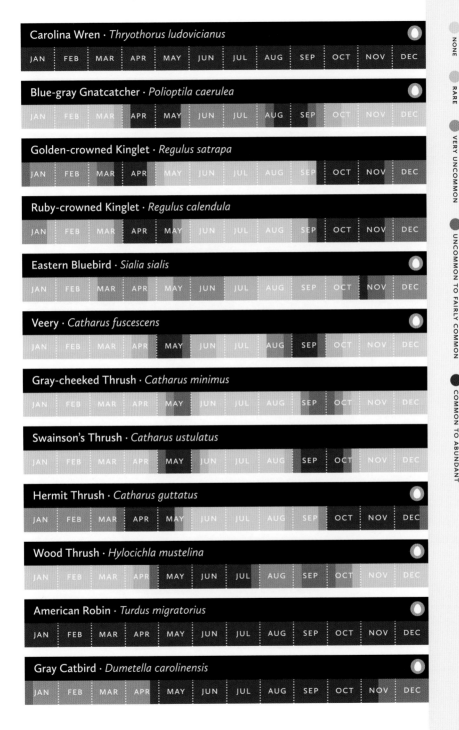

Carolina Wren · *Thryothorus ludovicianus*

| JAN | FEB | MAR | APR | MAY | JUN | JUL | AUG | SEP | OCT | NOV | DEC |

Blue-gray Gnatcatcher · *Polioptila caerulea*

| JAN | FEB | MAR | APR | MAY | JUN | JUL | AUG | SEP | OCT | NOV | DEC |

Golden-crowned Kinglet · *Regulus satrapa*

| JAN | FEB | MAR | APR | MAY | JUN | JUL | AUG | SEP | OCT | NOV | DEC |

Ruby-crowned Kinglet · *Regulus calendula*

| JAN | FEB | MAR | APR | MAY | JUN | JUL | AUG | SEP | OCT | NOV | DEC |

Eastern Bluebird · *Sialia sialis*

| JAN | FEB | MAR | APR | MAY | JUN | JUL | AUG | SEP | OCT | NOV | DEC |

Veery · *Catharus fuscescens*

| JAN | FEB | MAR | APR | MAY | JUN | JUL | AUG | SEP | OCT | NOV | DEC |

Gray-cheeked Thrush · *Catharus minimus*

| JAN | FEB | MAR | APR | MAY | JUN | JUL | AUG | SEP | OCT | NOV | DEC |

Swainson's Thrush · *Catharus ustulatus*

| JAN | FEB | MAR | APR | MAY | JUN | JUL | AUG | SEP | OCT | NOV | DEC |

Hermit Thrush · *Catharus guttatus*

| JAN | FEB | MAR | APR | MAY | JUN | JUL | AUG | SEP | OCT | NOV | DEC |

Wood Thrush · *Hylocichla mustelina*

| JAN | FEB | MAR | APR | MAY | JUN | JUL | AUG | SEP | OCT | NOV | DEC |

American Robin · *Turdus migratorius*

| JAN | FEB | MAR | APR | MAY | JUN | JUL | AUG | SEP | OCT | NOV | DEC |

Gray Catbird · *Dumetella carolinensis*

| JAN | FEB | MAR | APR | MAY | JUN | JUL | AUG | SEP | OCT | NOV | DEC |

BREEDS IN NEW YORK CITY AND ON LONG ISLAND

NONE

RARE

VERY UNCOMMON

UNCOMMON TO FAIRLY COMMON

COMMON TO ABUNDANT

Brown Thrasher · *Toxostoma rufum*

| JAN | FEB | MAR | APR | MAY | JUN | JUL | AUG | SEP | OCT | NOV | DEC |

Northern Mockingbird · *Mimus polyglottos*

| JAN | FEB | MAR | APR | MAY | JUN | JUL | AUG | SEP | OCT | NOV | DEC |

European Starling · *Sturnus vulgaris*

| JAN | FEB | MAR | APR | MAY | JUN | JUL | AUG | SEP | OCT | NOV | DEC |

American Pipit · *Anthus rubescens*

| JAN | FEB | MAR | APR | MAY | JUN | JUL | AUG | SEP | OCT | NOV | DEC |

Cedar Waxwing · *Bombycilla cedrorum*

| JAN | FEB | MAR | APR | MAY | JUN | JUL | AUG | SEP | OCT | NOV | DEC |

Snow Bunting · *Plectrophenax nivalis*

| JAN | FEB | MAR | APR | MAY | JUN | JUL | AUG | SEP | OCT | NOV | DEC |

Ovenbird · *Seiurus aurocapilla*

| JAN | FEB | MAR | APR | MAY | JUN | JUL | AUG | SEP | OCT | NOV | DEC |

Worm-eating Warbler · *Helmitheros vermivorum*

| JAN | FEB | MAR | APR | MAY | JUN | JUL | AUG | SEP | OCT | NOV | DEC |

Louisiana Waterthrush · *Parkesia motacilla*

| JAN | FEB | MAR | APR | MAY | JUN | JUL | AUG | SEP | OCT | NOV | DEC |

Rarely breeds here

Northern Waterthrush · *Parkesia noveboracensis*

| JAN | FEB | MAR | APR | MAY | JUN | JUL | AUG | SEP | OCT | NOV | DEC |

Blue-winged Warbler · *Vermivora cyanoptera*

| JAN | FEB | MAR | APR | MAY | JUN | JUL | AUG | SEP | OCT | NOV | DEC |

Black-and-white Warbler · *Mniotilta varia*

| JAN | FEB | MAR | APR | MAY | JUN | JUL | AUG | SEP | OCT | NOV | DEC |

Prothonotary Warbler · *Protonotaria citrea*

JAN	FEB	MAR	APR	MAY	JUN	JUL	AUG	SEP	OCT	NOV	DEC

Tennessee Warbler · *Oreothlypis peregrina*

JAN	FEB	MAR	APR	MAY	JUN	JUL	AUG	SEP	OCT	NOV	DEC

Orange-crowned Warbler · *Oreothlypis celata*

JAN	FEB	MAR	APR	MAY	JUN	JUL	AUG	SEP	OCT	NOV	DEC

Nashville Warbler · *Oreothlypis ruficapilla*

JAN	FEB	MAR	APR	MAY	JUN	JUL	AUG	SEP	OCT	NOV	DEC

Connecticut Warbler · *Oporornis agilis*

JAN	FEB	MAR	APR	MAY	JUN	JUL	AUG	SEP	OCT	NOV	DEC

Mourning Warbler · *Geothlypis philadelphia*

JAN	FEB	MAR	APR	MAY	JUN	JUL	AUG	SEP	OCT	NOV	DEC

Kentucky Warbler · *Geothlypis formosa*

JAN	FEB	MAR	APR	MAY	JUN	JUL	AUG	SEP	OCT	NOV	DEC

Common Yellowthroat · *Geothlypis trichas*

JAN	FEB	MAR	APR	MAY	JUN	JUL	AUG	SEP	OCT	NOV	DEC

Hooded Warbler · *Setophaga citrina*

JAN	FEB	MAR	APR	MAY	JUN	JUL	AUG	SEP	OCT	NOV	DEC

American Redstart · *Setophaga ruticilla*

JAN	FEB	MAR	APR	MAY	JUN	JUL	AUG	SEP	OCT	NOV	DEC

Cape May Warbler · *Setophaga tigrina*

JAN	FEB	MAR	APR	MAY	JUN	JUL	AUG	SEP	OCT	NOV	DEC

Northern Parula · *Setophaga americana*

JAN	FEB	MAR	APR	MAY	JUN	JUL	AUG	SEP	OCT	NOV	DEC

BREEDS IN NEW YORK CITY AND ON LONG ISLAND

NONE

RARE

VERY UNCOMMON

UNCOMMON TO FAIRLY COMMON

COMMON TO ABUNDANT

Magnolia Warbler · *Setophaga magnolia*

JAN	FEB	MAR	APR	MAY	JUN	JUL	AUG	SEP	OCT	NOV	DEC

Bay-breasted Warbler · *Setophaga castanea*

JAN	FEB	MAR	APR	MAY	JUN	JUL	AUG	SEP	OCT	NOV	DEC

Blackburnian Warbler · *Setophaga fusca*

JAN	FEB	MAR	APR	MAY	JUN	JUL	AUG	SEP	OCT	NOV	DEC

Yellow Warbler · *Setophaga petechia*

JAN	FEB	MAR	APR	MAY	JUN	JUL	AUG	SEP	OCT	NOV	DEC

Chestnut-sided Warbler · *Setophaga pensylvanica*

JAN	FEB	MAR	APR	MAY	JUN	JUL	AUG	SEP	OCT	NOV	DEC

Blackpoll Warbler · *Setophaga striata*

JAN	FEB	MAR	APR	MAY	JUN	JUL	AUG	SEP	OCT	NOV	DEC

Black-throated Blue Warbler · *Setophaga caerulescens*

JAN	FEB	MAR	APR	MAY	JUN	JUL	AUG	SEP	OCT	NOV	DEC

Palm Warbler · *Setophaga palmarum*

JAN	FEB	MAR	APR	MAY	JUN	JUL	AUG	SEP	OCT	NOV	DEC

Pine Warbler · *Setophaga pinus*

JAN	FEB	MAR	APR	MAY	JUN	JUL	AUG	SEP	OCT	NOV	DEC

Yellow-rumped Warbler · *Setophaga coronata*

JAN	FEB	MAR	APR	MAY	JUN	JUL	AUG	SEP	OCT	NOV	DEC

Yellow-throated Warbler · *Setophaga dominica*

JAN	FEB	MAR	APR	MAY	JUN	JUL	AUG	SEP	OCT	NOV	DEC

Prairie Warbler · *Setophaga discolor*

JAN	FEB	MAR	APR	MAY	JUN	JUL	AUG	SEP	OCT	NOV	DEC

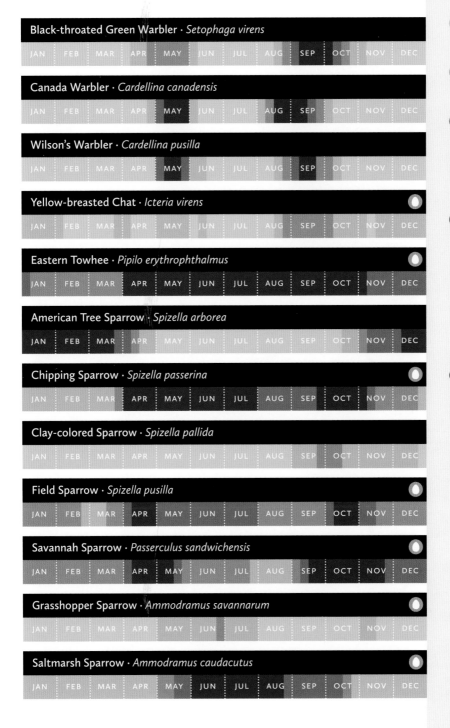

Black-throated Green Warbler · *Setophaga virens*

JAN FEB MAR APR MAY JUN JUL AUG SEP OCT NOV DEC

Canada Warbler · *Cardellina canadensis*

JAN FEB MAR APR MAY JUN JUL AUG SEP OCT NOV DEC

Wilson's Warbler · *Cardellina pusilla*

JAN FEB MAR APR MAY JUN JUL AUG SEP OCT NOV DEC

Yellow-breasted Chat · *Icteria virens*

JAN FEB MAR APR MAY JUN JUL AUG SEP OCT NOV DEC

Eastern Towhee · *Pipilo erythrophthalmus*

JAN FEB MAR APR MAY JUN JUL AUG SEP OCT NOV DEC

American Tree Sparrow · *Spizella arborea*

JAN FEB MAR APR MAY JUN JUL AUG SEP OCT NOV DEC

Chipping Sparrow · *Spizella passerina*

JAN FEB MAR APR MAY JUN JUL AUG SEP OCT NOV DEC

Clay-colored Sparrow · *Spizella pallida*

JAN FEB MAR APR MAY JUN JUL AUG SEP OCT NOV DEC

Field Sparrow · *Spizella pusilla*

JAN FEB MAR APR MAY JUN JUL AUG SEP OCT NOV DEC

Savannah Sparrow · *Passerculus sandwichensis*

JAN FEB MAR APR MAY JUN JUL AUG SEP OCT NOV DEC

Grasshopper Sparrow · *Ammodramus savannarum*

JAN FEB MAR APR MAY JUN JUL AUG SEP OCT NOV DEC

Saltmarsh Sparrow · *Ammodramus caudacutus*

JAN FEB MAR APR MAY JUN JUL AUG SEP OCT NOV DEC

NONE
RARE
VERY UNCOMMON
UNCOMMON TO FAIRLY COMMON
COMMON TO ABUNDANT

BREEDS IN NEW YORK CITY AND ON LONG ISLAND

Seaside Sparrow · *Ammodramus maritimus* ⬤

JAN	FEB	MAR	APR	MAY	JUN	JUL	AUG	SEP	OCT	NOV	DEC

Fox Sparrow · *Passerella iliaca*

JAN	FEB	MAR	APR	MAY	JUN	JUL	AUG	SEP	OCT	NOV	DEC

Song Sparrow · *Melospiza melodia* ⬤

JAN	FEB	MAR	APR	MAY	JUN	JUL	AUG	SEP	OCT	NOV	DEC

Lincoln's Sparrow · *Melospiza lincolnii*

JAN	FEB	MAR	APR	MAY	JUN	JUL	AUG	SEP	OCT	NOV	DEC

Swamp Sparrow · *Melospiza georgiana* ⬤

JAN	FEB	MAR	APR	MAY	JUN	JUL	AUG	SEP	OCT	NOV	DEC

White-throated Sparrow · *Zonotrichia albicollis*

JAN	FEB	MAR	APR	MAY	JUN	JUL	AUG	SEP	OCT	NOV	DEC

White-crowned Sparrow · *Zonotrichia leucophrys*

JAN	FEB	MAR	APR	MAY	JUN	JUL	AUG	SEP	OCT	NOV	DEC

Dark-eyed Junco · *Junco hyemalis* ⬤

JAN	FEB	MAR	APR	MAY	JUN	JUL	AUG	SEP	OCT	NOV	DEC

Summer Tanager · *Piranga rubra*

JAN	FEB	MAR	APR	MAY	JUN	JUL	AUG	SEP	OCT	NOV	DEC

Scarlet Tanager · *Piranga olivacea* ⬤

JAN	FEB	MAR	APR	MAY	JUN	JUL	AUG	SEP	OCT	NOV	DEC

Northern Cardinal · *Cardinalis cardinalis* ⬤

JAN	FEB	MAR	APR	MAY	JUN	JUL	AUG	SEP	OCT	NOV	DEC

Rose-breasted Grosbeak · *Pheucticus ludovicianus* ⬤

JAN	FEB	MAR	APR	MAY	JUN	JUL	AUG	SEP	OCT	NOV	DEC

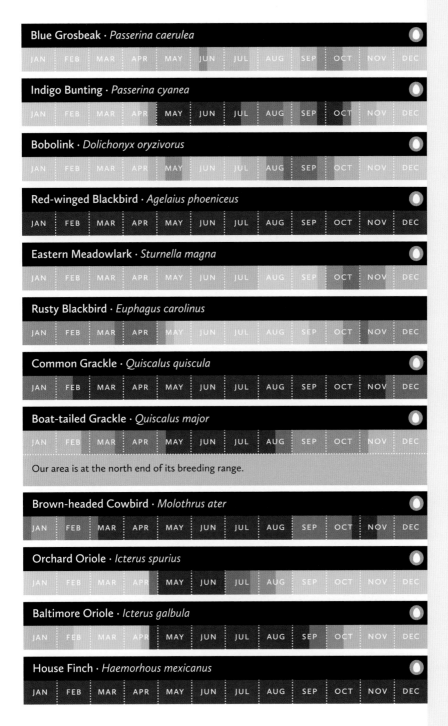

Blue Grosbeak · *Passerina caerulea*

JAN	FEB	MAR	APR	MAY	JUN	JUL	AUG	SEP	OCT	NOV	DEC

Indigo Bunting · *Passerina cyanea*

JAN	FEB	MAR	APR	MAY	JUN	JUL	AUG	SEP	OCT	NOV	DEC

Bobolink · *Dolichonyx oryzivorus*

JAN	FEB	MAR	APR	MAY	JUN	JUL	AUG	SEP	OCT	NOV	DEC

Red-winged Blackbird · *Agelaius phoeniceus*

JAN	FEB	MAR	APR	MAY	JUN	JUL	AUG	SEP	OCT	NOV	DEC

Eastern Meadowlark · *Sturnella magna*

JAN	FEB	MAR	APR	MAY	JUN	JUL	AUG	SEP	OCT	NOV	DEC

Rusty Blackbird · *Euphagus carolinus*

JAN	FEB	MAR	APR	MAY	JUN	JUL	AUG	SEP	OCT	NOV	DEC

Common Grackle · *Quiscalus quiscula*

JAN	FEB	MAR	APR	MAY	JUN	JUL	AUG	SEP	OCT	NOV	DEC

Boat-tailed Grackle · *Quiscalus major*

JAN	FEB	MAR	APR	MAY	JUN	JUL	AUG	SEP	OCT	NOV	DEC

Our area is at the north end of its breeding range.

Brown-headed Cowbird · *Molothrus ater*

JAN	FEB	MAR	APR	MAY	JUN	JUL	AUG	SEP	OCT	NOV	DEC

Orchard Oriole · *Icterus spurius*

JAN	FEB	MAR	APR	MAY	JUN	JUL	AUG	SEP	OCT	NOV	DEC

Baltimore Oriole · *Icterus galbula*

JAN	FEB	MAR	APR	MAY	JUN	JUL	AUG	SEP	OCT	NOV	DEC

House Finch · *Haemorhous mexicanus*

JAN	FEB	MAR	APR	MAY	JUN	JUL	AUG	SEP	OCT	NOV	DEC

BREEDS IN NEW YORK CITY AND ON LONG ISLAND

NONE

RARE

VERY UNCOMMON

UNCOMMON TO FAIRLY COMMON

COMMON TO ABUNDANT

Purple Finch · *Haemorhous purpureus*

JAN	FEB	MAR	APR	MAY	JUN	JUL	AUG	SEP	OCT	NOV	DEC

Pine Siskin · *Spinus pinus*

JAN	FEB	MAR	APR	MAY	JUN	JUL	AUG	SEP	OCT	NOV	DEC

American Goldfinch · *Spinus tristis*

JAN	FEB	MAR	APR	MAY	JUN	JUL	AUG	SEP	OCT	NOV	DEC

House Sparrow · *Passer domesticus*

JAN	FEB	MAR	APR	MAY	JUN	JUL	AUG	SEP	OCT	NOV	DEC

Rarities

Rarities are birds that occur in this area a few times annually to less than annually and at least once every ten years.

Ross's Goose, *Chen rossii*

Tufted Duck, *Aythya fuligula*—reliable, though unofficial, reports from Long Island becoming more frequent

Pacific Loon, *Gavia pacifica*

Western Grebe, *Aechmophorus occidentalis*

Black-capped Petrel, *Pterodroma hasitata*—pelagic; there is one non-pelagic record relating to Hurricane Irene in 2011

Fea's Petrel, *Pterodroma feae*—pelagic

Great Shearwater, *Puffinus gravis*—pelagic; sometimes seen in coastal environments

Manx Shearwater, *Puffinus puffinus*—pelagic; sometimes seen in coastal environments

White-faced Storm Petrel, *Pelagodroma marina*—pelagic

Leach's Storm Petrel, *Oceanodroma leucorhoa*—pelagic; rarely seen in coastal environments

Band-rumped Storm Petrel, *Oceanodroma castro*—pelagic

Magnificent Frigatebird, *Fregata magnificens*—sometimes seen in coastal environments

Brown Pelican, *Pelecanus occidentalis*

Cattle Egret, *Bubulcus ibis*

White-faced Ibis, *Plegadis chihi*

Mississippi Kite, *Ictinia mississippiensis*

Northern Goshawk, *Accipiter gentilis*

Golden Eagle, *Aquila chrysaetos*—more likely in fall

Virginia Rail, *Rallus limicola*

Black-necked Stilt, *Himantopus mexicanus*—most sightings occur in May

Wilson's Plover, *Charadrius wilsonia*

Upland Sandpiper, *Bartramia longicauda*—rare and breeds rarely here

Bar-tailed Godwit, *Limosa lapponica*

Ruff, *Calidris pugnax*

Curlew Sandpiper, *Calidris ferruginea*

Red-necked Stint, *Calidris ruficollis*

Red-necked Phalarope

Red-necked Phalarope, *Phalaropus lobatus*

Red Phalarope, *Phalaropus fulicarius*

Pomarine Jaeger, *Stercorarius pomarinus*—occasional coastal reports

Parasitic Jaeger, *Stercorarius parasiticus*—most often seen from the South Shore of Long Island during migration

Dovekie, *Alle alle*

Common Murre, *Uria aalge*—pelagic; coastal reports are very rare

Thick-billed Murre, *Uria lomvia*

Black Guillemot, *Cepphus grylle*

Atlantic Puffin, *Fratercula arctica*—pelagic

Black-legged Kittiwake, *Rissa tridactyla*

Black-headed Gull, *Chroicocephalus ridibundus*

Little Gull, *Hydrocoloeus minutus*

Franklin's Gull, *Leucophaeus pipixcan*

Mew Gull, *Larus canus*

Arctic Tern, Sterna *paradisaea*

Sandwich Tern, *Thalasseus sandvicensis*

Barn Owl, *Tyto alba*—rare, but it does breed in the area

Barred Owl, *Strix varia*—rare, but it does breed here

Chuck-will's-widow, *Antrostomus carolinensis*

Rufous Hummingbird, *Selasphorus rufus*

Gyrfalcon, *Falco rusticolus*

Alder Flycatcher, *Empidonax alnorum*—rarely breeds here

Ash-throated Flycatcher, *Myiarchus cinerascens*

Western Kingbird, *Tyrannus verticalis*

Northern Shrike, *Lanius excubitor*

Cliff Swallow, *Petrochelidon pyrrhonota*

Cave Swallow, *Petrochelidon fulva*—rare fall visitor

Varied Thrush, *Ixoreus naevius*

Lapland Longspur, *Calcarius lapponicus*

Golden-winged Warbler, *Vermivora chrysoptera*

Cerulean Warbler, *Setophaga cerulea*—a rare spring visitor more common in the western part of the area

Black-throated Gray Warbler, *Setophaga nigrescens*

Townsend's Warbler, *Setophaga townsendi*

Vesper Sparrow, *Pooecetes gramineus*—very rarely a breeder here

Lark Sparrow, *Chondestes grammacus*

Lark Bunting, *Calamospiza melanocorys*—only five official records; the most recent was at Jones Beach in fall 2008

Le Conte's Sparrow, *Ammodramus leconteii*

Nelson's Sparrow, *Ammodramus nelsoni*

Western Tanager, *Piranga ludoviciana*

Painted Bunting, *Passerina ciris*

Dickcissel, *Spiza americana*

Yellow-headed Blackbird, *Xanthocephalus xanthocephalus*

Evening Grosbeak, *Coccothraustes vespertinus*

White-winged Dove, *Zenaida asiatica*

Red Crossbill, *Loxia curvirostra*—irruptive; 2012 was probably the best year in recent memory

White-winged Crossbill, *Loxia leucoptera*—irruptive; occurrence patterns similar to Red Crossbill

Common Redpoll, *Acanthis flammea*—irruptive; a slightly steadier presence than Red Crossbill and White-winged Crossbill

Accidentals

These birds have occurred in the area fewer than ten times between 2000 and 2014 and are considered accidental or vagrant species.

Black-bellied Whistling-Duck, *Dendrocygna autumnalis*—one official report since 2000, in 2010 in Queens

Fulvous Whistling-Duck, *Dendrocygna bicolor*—one official report since 2000, in 2006 in Queens

Trumpeter Swan, *Cygnus buccinators*—there was an overwintering pair at Upper Lake, near Yaphank in Suffolk County, from winter 2009-10 until one was shot in January of 2013; the remaining swan did not return the following year

Northern Fulmar, *Fulmarus glacialis*—pelagic; coastal reports are very rare

White-tailed Tropicbird, *Phaethon lepturus*—all recent reports are associated with 2011's Hurricane Irene

Brown Booby, *Sula leucogaster*—no official records since 1997

Western Reef-Heron, *Egretta gularis*—one official record in 2009, Coney Island Creek

White Ibis, *Eudocimus albus*—several New York State records in 2011 but only one for this area, at Mount Loretto, Staten Island

Swallow-tailed Kite, *Elanoides forficatus*

Yellow Rail, *Coturnicops noveboracensis*—two reports in the area since 2000

Black Rail, *Laterallus jamaicensis*—a few sightings by reliable reporters, but no official records

King Rail, *Rallus elegans*—two reports, 2006 and 2013

Purple Gallinule, *Porphyrio martinicus*—one record since 1978, Prospect Park in 2004

Common Gallinule, *Gallinula galeata*

Sandhill Crane, *Grus Canadensis*—a few reported sightings but no official records in the area since 1995

Northern Lapwing, *Vanellus vanellus*

Sharp-tailed Sandpiper, *Calidris acuminata*—all reported sightings from Jamaica Bay NWR

Little Stint, *Calidris minuta*—no records since 2000

Great Skua, *Stercorarius skua*—one record, from Montauk in 2002

South Polar Skua, *Stercorarius maccormicki*—three coastal records since 2001

Long-tailed Jaeger, *Stercorarius longicaudus*—a rare pelagic bird; the only coastal reports are from 2012 and 2013

Ivory Gull, *Pagophila eburnea*

Sabine's Gull, *Xema sabini*

Gray-hooded Gull, *Chroicocephalus cirrocephalus*—one record, Coney Island 2011

Black-tailed Gull, *Larus crassirostris*—two official records in New York State beginning in 1999, both at Jones Beach

Western Gull, *Larus occidentalis*—one pelagic record, 2006

Thayer's Gull, *Larus thayeri*

Sooty Tern, *Onychoprion fuscatus*—most records in this area are from 2011 and associated with Hurricane Irene

Bridled Tern, *Onychoprion anaethetus*

Elegant Tern, *Thalasseus elegans*—all reported sightings are from July 2013

Band-tailed Pigeon, *Patagioenas fasciata*—one official record, Brookhaven, 2008

Eurasian Collared-Dove, *Streptopelia decaocto*

Common Ground-Dove, *Columbina passerina*—Captree State Park in 2010 and Jones Beach, 2014

Boreal Owl, *Aegolius funereus*—one official record in the area, Central Park, 2004

Calliope Hummingbird, *Selasphorus calliope*

Pileated Woodpecker, *Dryocopus pileatus*—though common in New York State, it is almost unheard of in the area

Hammond's Flycatcher, *Empidonax hammondii*—one official record, Jones Beach, 2001

Say's Phoebe, *Sayornis saya*

Couch's Kingbird, *Tyrannus couchii*—the only sightings in the area were in Manhattan's West Village, December 2014

Cassin's Kingbird, *Tyrannus vociferans*—official records from 2007 only, but numerous reports of one at Floyd Bennett Field November 2014 to January 2015

Gray Kingbird, *Tyrannus dominicensis*—two official records in the area beginning in 1989

Scissor-tailed Flycatcher, *Tyrannus forficatus*

Fork-tailed Flycatcher, *Tyrannus savana*

Loggerhead Shrike, *Lanius ludovicianus*

Bell's Vireo, *Vireo bellii*—a difficult bird to identify securely; the last accepted report for New York State was in 1996

Sedge Wren, *Cistothorus platensis*—only two recorded observations; no official records

Northern Wheatear, *Oenanthe oenanthe*

Mountain Bluebird, *Sialia currucoides*

Townsend's Solitaire, *Myadestes townsendi*

Bicknell's Thrush, *Catharus bicknelli*—because this thrush is so difficult to identify with respect to its more common relatives, the Gray-cheeked, Swainson's, and Hermit Thrushes, few observations are accepted

Eastern Yellow Wagtail, *Motacilla tschutschensis/flava*—one official record, Plumb Beach, 2008

Bohemian Waxwing, *Bombycilla garrulus*

Smith's Longspur, *Calcarius pictus*—one official record, Jones Beach, 2007

Swainson's Warbler, *Limnothlypis swainsonii*—accepted reports in 2004 and 2005

Virginia's Warbler, *Oreothlypis virginiae*—one occurrence, Alley Pond Park, fall 2012

MacGillivray's Warbler, *Geothlypis tolmiei*—two records for this area, 1999 on Staten Island and 2004 in Forest Park

Grace's Warbler, *Setophaga graciae*—one accepted record, Point Lookout 2012

Hermit Warbler, *Setophaga occidentalis*—one official record, Sunken Meadow State Park, 2010

Cassin's Sparrow, *Peucaea cassinii*—one official record, Jones Beach, 2000

Harris's Sparrow, *Zonotrichia querula*—one official record since 2000, during a Christmas bird count, January 2005

Golden-crowned Sparrow, *Zonotrichia atricapilla*—three official records, the last in 1995

Brewer's Blackbird, *Euphagus cyanocephalus*—three reports since 2000

Scott's Oriole, *Icterus parisorum*—one official record, New York City's Union Square Park, winter 2007–8

Bibliography

PRIMARY SOURCES

Barton, Howard, and Patricia I. Pelkowski. *A Seasonal Guide to Bird Finding on Long Island.* Smithtown, NY: ECSS, 1999.

Drennan, Susan Roney. *Where to Find Birds in New York State: The Top 500 Sites.* Syracuse, NY: Syracuse University Press, 1981.

Fowle, Marcia T., and Paul Kerlinger. *The New York City Audubon Society Guide to Finding Birds in the Metropolitan Area.* Ithaca, NY: Comstock Publishing Associates, Cornell University Press, 2001.

OTHER SOURCES

Albright, Rodney, and Priscilla Albright. *Short Nature Walks on Long Island.* 3rd ed. Chester, CT: Globe Pequot Press, 2001.

AOU Checklist of North American Birds. 7th ed., 54th Supplement. September 2013.

Berenson, Richard J., and Raymond Carroll. *Barnes & Noble Complete Illustrated Map and Guidebook to Central Park.* New York: Silver Lining Books / Berenson Design & Books, 2003.

Bird, Christiane. *New York State.* 3rd ed. Emeryville, CA: Avalon Travel, 2003.

Burger, Michael F. *Important Bird Areas of New York.* 2nd ed. Albany: Audubon New York, 2005.

Card, Skip. *Moon Take a Hike: New York City: 80 Hikes within 2 Hours of Manhattan.* Berkeley, CA: Avalon Travel, 2012.

Case, Dan. *AMC's Best Day Hikes near New York City: Four-Season Guide to 50 of the Best Trails in New York, Connecticut, and New Jersey.* Boston: Appalachian Mountain Club Books, 2010.

Day, Leslie. *Field Guide to the Natural World of New York City.* Baltimore: Johns Hopkins University Press, 2007.

Deutsch, Alice. *Natural History of New York City's Parks and Great Gull Island.* New York: Linnaean Society of New York, 2007.

Dunne, Pete, and David Sibley. *Hawks in Flight: The Flight Identification of North American Migrant Raptors.* Boston: Houghton Mifflin, 1988.

Kaufman, Kenn. *City Birding: True Tales of Birds and Birdwatching in Unexpected Places.* Mechanicsburg, PA: Stackpole Books, 2003.

Knowler, Donald. *The Falconer of Central Park.* New York: Karz-Cohl, 1984.

Lancaster, Clay. *Prospect Park Handbook.* New York: W. H. Rawls, 1967.

Masterson, Eric A. *Birdwatching in New Hampshire.* Lebanon, NH: University Press of New England, 2013.

McGowan, Kevin J. *The Second Atlas of Breeding Birds in New York State.* Ithaca, NY: Comstock Publishing Associates, Cornell University Press, 2008.

Miller, Ian. "New York City and Long Island: 14th to 21st January 2012." http://www.surfbirds.com/trip_report.php?id=2162.

Peterson, Roger Tory. *Peterson Field Guide to Birds of North America.* Boston: Houghton Mifflin, 2008.

Sanderson, Eric W. *Mannahatta: A Natural History of New York City.* New York: Abrams, 2009.

Sibley, David. *The Sibley Field Guide to Birds of Eastern North America.* New York: Alfred A. Knopf, 2003.

Stephenson, Tom, and Scott Whittle. *Warbler Guide.* Princeton, NJ: Princeton University Press, 2013.

Stokes, Donald W., and Lillian Q. Stokes. *Stokes Beginner's Guide to Shorebirds.* Boston: Little, Brown, 2001.

Thompson, John Henry. *Geography of New York State.* Syracuse, NY: Syracuse University Press, 1977.

Tove, Michael H. *Guide to the Offshore Wildlife of the Northern Atlantic.* Austin: University of Texas Press, 2000.

Turner, John L. "Coastal and Pelagic Birds of Long Island." http://www.cresli.org/cresli/Birds/LIbirds.html.

VanDiver, Bradford B. *Roadside Geology of New York.* Missoula, MT: Mountain Press Publishers, 1985.

Winn, Marie. *Central Park in the Dark: More Mysteries of Urban Wildlife.* New York: Farrar, Straus and Giroux, 2008.

———. *Red-Tails in Love: A Wildlife Drama in Central Park.* New York: Pantheon Books, 1998.

WEBSITES

American Birding Association, www.aba.org

Audubon New York, ny.audubon.org/

The Battery Conservancy, www.thebattery.org/

Brooklyn Bird Club, www.brooklynbirdclub.org/

Cornell's Lab of Ornithology, www.eBird.org

Cornell University, NYSBirds-L, Listowner: Chris Tessaglia-Hymes, www.north-
eastbirding.com/NYSbirdsWELCOME

Linnaean Society of New York, www.Linnaeannewyork.org

Metropolitan Transit Authority, www.mta.info/

National Park Service, www.nps.gov/index.htm

Nature Conservancy, www.tnc.org

New York City Audubon, www.nycaudubon.org

New York City Department of Parks & Recreation, www.nycgovparks.org

New York City Parks, www.nycgovparks.org/

New York Harbor Parks, www.nyharborparks.org/index.html

New York Important Bird Areas, National Audubon Society, netapp.audubon.org
/iba

New York State Avian Records Committee, www.nybirds.org/NYSARC/

New York State Parks, www.nysparks.com/parks/

North Fork Audubon, www.northforkaudubon.org/

Queens County Bird Club, www.qcbirdclub.org/

Randall's Island Park Alliance, www.randallsisland.org/

BLOGS AND PERSONAL WEBSITES

Andrew Baksh's Birding Dude blog, http://birdingdude.blogspot.com/

Debbie Becker's www.birdingaroundnyc.com

Jack Rothman's, www.cityislandbirds.com/City_Island_Birds.html

Long Island Birding.com, www.Libirding.com

Marie Winn's Central Park Nature News, www.mariewinnnaturenews.blogspot
.com

Paul Guris's See Life Paulagics, www.paulagics.com

Phil Jeffrey's Central Park Birding, www.philjeffrey.net/cpb_index.html

Rob Jett's the City Birder, www.citybirder.blogspot.com/

10,000 Birds, www.10000birds.com/

Index

Page numbers in *italics* refer to maps and photographs.

Acme Pond, 139, 151–54
Alley Pond, 113
Alley Pond Park, 6, 109–15, *110*
Allison Pond Park, 139, 170–71
Alpine Mountain, NJ, 7
American Birding Association, birding ethics, 11–13
American Veterans Memorial Pier, 75, 83
APEC, 109, 111–12, 114
Arbutus Pond, 150
Arshamomaque Pond Preserve, 256
Arshamomaque Preserve, 256
Arverne Piping Plover Nesting Area, 103, 116
Atlantic Flyway, 1, 17
Audubon, National Society, 9
Audubon, New York, 9, 225
Audubon, New York City, 9, 54
Avocet, American, *89*, 91

Bailey Arboretum, 178, 209
Baisley Pond Park, 87, 115–16
Battery Harris East, 97
Battery, the, 48–49
Bayard Cutting Arboretum State Park, 237
Best of Chart, 6

Betts Creek, 81
Big Egg Marsh, aka Broad Channel American Park, 93–94
birding by month, 2–4
birding ethics. *See* American Birding Association, birding ethics
Bittern, Least, 167, 247
Blackbird, Red-winged, 37, 41, 68, 91, 97, 103, 109, 111, 114, 125–26, 153, 160, 169, 174, 188, 204, 211, 227, 234, 244
Blackbird, Rusty, 27, 112, 113, 164–65, 172, 197
Bluebelt, Staten Island, 139
Bluebird, Eastern, 27, 67, 211, 212, 227, 236, 252
Blue Heron Park Preserve, 7, 139, 149–51
Blydenburgh Park, 232, 234
Bobolink, 71, 103, 158, 212, 216, 218, 227
Bobwhite, Northern, 235
Brant, 44, 47, 53, 74, 77, 80–81, 91, 126, 160, 185, 188, 191, 234, 247
Breezy Point, 6, 85, 99–102
Bridge Creek, 168–70
Bronx, *120*, 120–138
Bronx Kill, 40
Bronx Zoo, 127–31
Brooklyn, 57–84, *58*

Brooklyn Army Terminal Pier 4, 75
Brooklyn Botanic Garden, 57, 59, 62–65
Brooklyn Bridge Park, 74, 82
Brooklyn Coastal Winter Waterfowl
    Viewing, 6, 73
Bryant Park, 9, 17, 50
Buffalo Farm, the, 242
Bufflehead, 24, 27, 44, 46–47, 71, 74–77,
    80–81, 91, 103, 113, 118, 126, 148, 150–
    51, 174, 198, 199, 246–47, 251, 258
Bunting, Indigo, 21, 26, 60, 117, 126, 157,
    158, 160, 176, 204, 217–18, 224, 232
Bunting, Snow, 41, 70–71, 77, 95, 96, 97,
    100, 145, 148, 172, 181, 183, 209, 218
Bush Terminal Piers Park, 74–75, 83

Calverton Grasslands, 241
Calverton Ponds Preserve, 242
Calvert Vaux Park, 76, 83
Camman's Pond Park, 178, 201–2
Camp Hero State Park, 258, 260, 262
Canarsie Park, 80
Canarsie Pier, 80, 93
Canvasback, 27, 34, 46, 91, 118, 126, 229,
    236, 247
Captree State Park, 198, 224–25
Cardinal, Northern, i, 27, 205
Carl Schurz Park, 53
Catbird, Gray, vi, 26, 112
Caumsett State Historic Park, 207, 215,
    225–28
Cedar Beach County Park, 257
Cemetery of the Evergreens, 119
Central Park, 2–3, 6, 17, 19–29, 21; Azalea
    Pond, 20; Belvedere Castle, 7, 21, 22,
    27; Block House, 24; Bow Bridge, 21;
    Conservatory Garden, 26, 28; Dawn
    Redwood Trees, 23; Evodia Field,
    20; Falconer's Statue, 23; the Gill,
    20; Great Hill, 24, 28; Hallett Nature
    Sanctuary, 25; the Lake, 21, 22, 24,
    27; Laupots Bridge, 20, 29; Lily Pool,
    24; the Loch, 24; Locust Grove, 25;
    Maintenance Meadow, 20, 22, 28, 29;
    North Woods, 19, 20, 24, 25, 28; Oak
    Bridge, 21; the Oven, 20, 29; Pinetum,
    25; the Point, 20, 29; Polish Statue,
    22; the Pond, 25, 27; the Pool, 24;
    Ramble, 19–22, 22, 28–29; the Ravine,
    24; Reservoir, 19, 20, 24, 25, 27, 29;
    the Riviera, 21; Shakespeare's Garden,
    21, 28; Sparrow Rock, 24; Strawberry
    Fields, 19–20, 21, 22–23; Summit
    Rock, 24; Tanner's Spring, 24, 29;
    Triplets Bridge, 24; Tupelo Meadow,
    20; Turtle Pond, 21–22; Upper Lobe,
    21; Wagner's Cove, 21; Willow Rock,
    20, 21
Centre Island Town Park, 209
Charles E. Ransom Beach, 209
Chickadee, Black-capped, 27, 34, 39, 64,
    67, 161, 188, 196, 205
Clay Pit Ponds State Park Preserve, 139,
    162–64
Cloisters, the, 33, 35–37, 46
Clove Lakes Park, 6, 139, 141–45, 142
Coney Island Creek, 76–77
Coney Island Creek Park, 77
Coney Island Pier, 77
Conference House Park, 6, 139, 160–62,
    161; South Pole, 160
Connetquot River State Park, 235–37
Coot, American, 144, 199
Cormorant, Double-crested, 188
Cormorant, Great, 41, 71, 103, 126, 148,
    158, 174, 247, 261
Cow Meadow Park and Preserve, 178,
    188–89, 201
Creeper, Brown, 27, 34, 60
Crossbill, Red-winged, 181, 183, 218

Crossbill, White-winged, 181, 183, 218
Crow, Fish, 71
Cuckoo, Black-billed, 91
Cuckoo, Yellow-billed, 158, 211, 236, 242
Cupsogue Beach County Park, 245, 247, 248–50

David Weld Sanctuary, 234
Dead Horse Bay, 72–73, 79–80
Dead Horse Point, 72–73, 79–80
Democrat Point, 7, 217
Denis and Catherine Krusos Research Area, 242
Dickcissel, 71, 216, 218
Ditch Plains, 263–264
Dove, Mourning, 18, 52, 63, 109, 147
Dovekie, 7
Dowitcher, Long-billed, 169, 197
Dowitcher, Short-billed, 169, 188, 249–50
Downs Farm Preserve, 257
Drier-Offerman Park. *See* Calvert Vaux Park
Drip, the, 52–53
Duck, American Black, 34, 37, 40, 41, 114, 134, 169, 199, 227, 234, 247
Duck, Harlequin, 183, *184*, 185, 261
Duck, Long-tailed, 74, 77, 97, 100, 147–48, 153, 155, 158, 173, 183, 185, 198, 207, 209, 211, 218, 221, 227, 229, 232, 239, 244, 246–47, 249, 253, 257, 261
Duck, Ring-necked, 27, 60, 114–15, 150, 165, 196, 199, 211, 234, 236, 242, 258
Duck, Ruddy, 27, 37, 60, 81, 91, 112, 114–15, 134, 175, 188, 196, 199
Duck, Tufted, 211
Duck, Wood, 3, 37, 60, 63, 71, 113–15, 118, 128, 134, 136, 144, 150–51, 158–59, 164–65, 167, *195*, 196, 199, 208, 210, 234–36, 242, 256

Dune Road, 245–48
Dunlin, 71, 91, 148, 191, *220*, 221, 232, 247, 260
Dyker Beach, 76

Eagle, Bald, 31, 33–34, 46, 55, 60, 137, 155, 158, 160–61, 174, 211, 218, 227, 252
Eagle, Golden, 174, 227
East River, 53
Edgemere Landfill, 85, 102–4
Egret, Great, 33, 39, 41, 106, 111, 114, 126, 134, 138, 143, 148, 153, 160, 169, 182, 188, 191, 206, 217, 231, 250
Egret, Snowy, 39, 41, 91, 103, 111, 114, 126, *147*, 148, 153, 169, 188, 217
Eider, Common, 183, 185, *216*, 244, 247, 255, 261
Eider, King, 243, 244, 261
Endangered Species Act, 8
EPCAL, 215, 241–42
Evergreen Cemetery, 119

Falcon, Peregrine, 2, 4, 8, 18, 34, 40, 41, 46, 53–55, 63–64, 69, 71, 97, 103, 148, 174, 182–83, 217–18, 227, 244, *246*, 247; nest cams, 56
Finch, House, 27, 34, 37, 147, 150
Finch, Purple, 27, 60, 91, 126, 143, 218, 232, 236, 252
Flicker, Northern, 27, 41, 43, 62–63, 67, 91, 97, 113, 125, 143, 151, 158, 188, 205, 262
Floyd Bennett Field, 6, 57, *69*, 69–72
Flushing Meadows Corona Park, 85, 117–18
Flycatcher, Acadian, 26, 60, 236
Flycatcher, Alder, 26, 236
Flycatcher, Great Crested, 26, 33, 60, 91, 106, 114, 143, 150, *194*, 196, 211, 252, 256
Flycatcher, Least, 26

Flycatcher, Olive-sided, 26
Flycatcher, Willow, 26, 83, 91, 103, 114, 125, 148, 182, 191, 217, 227, 232, 236, 251
Flycatcher, Yellow-bellied, 26
Forest Park, 6, 104–107, *105*
Fort Tilden, 6, 7, 85, 93–99, *94–95*, 101–2, 183; Battery Harris East, 97
Fort Tryon Park, 33, 35–36
Fort Wadsworth, 174–75
Fountain Avenue Landfill, 81
Four Sparrow Marsh, 84
Francis Ponds, Upper and Lower, 210
Fresh Creek Park, 81
Freshkills Park, 176–77

Gadwall, 24, 27, 37, 40, 41, 44, *56*, 74, 91, 114–15, 126, 151, 166–67, 169, 175, 196, 199, 236
Gallinule, Common, 169
Gannet, Northern, 74, 77, 97, 100, 147–48, 160–61, 181, 183, 185, 218, 221, 261
Garvies Point Preserve, 208
Gateway National Recreation Area, 57, 87, 145
Gerritsen Creek, 73, 77, 79
Gilgo Beach, 198, 224–25
Gnatcatcher, Blue-Gray, 26, 68, 91, 106, 113, 143, 194, 196, 205, 232
Godwit, Hudsonian, 91
Godwit, Marbled, 249
Goethals Pond, 168–170
Goethals Pond Complex, *167*, 168–70
Goldeneye, Barrow's, 207, 209
Goldeneye, Common, 74, 76, 91, 126, 148, 153, 155, 158, 160, 174, 183, 199, 209, 218, 221, 227, 229, 247, 257
Golden Triangle Sod Farms, 243
Goldfinch, American, 27, 34, 37, 60, 107, *108*, 109, 111, 114, 134, 147, 158, 169, 188, 227, 236

Goldsmith's Inlet Park, 257
Goose, Cackling, 134, 232, 243
Goose, Canada, 37, 74, 134, 196, 232, 247
Goose, Greater White-fronted, 134, 173, 232, 238
Goose, Pink-footed, 232
Goose, Ross's, 238
Goose, Snow, 68, 71, 91, 103, 134, 173, 238
Goshawk, Northern, 227
Governors Island, 18, 43–45
Grackle, Boat-tailed, 94, 103, 174, 182, 217, 247
Grackle, Common, 169
Grant Park Pond, 178, 199–200
Gravesend Bay, 75–76
Great Kills Park, 139, 145–49, *146*; Crooke's Point, 147
Grebe, Horned, 71, 74, 76, 80, 92, 97, 126, 148, 153, 155, 161, 174–75, 183, 185, 209, 218, 244, 247, 257
Grebe, Pied-billed, 24, 27, 63, 91–92, 112, 114, 175, 196, 199
Grebe, Red-necked, 41, 115, 148, 155, 174, 185, 261
Greenbelt, Long Island, 178, 196, 212, 233, 234
Greenbelt, Staten Island, 139, 164, 171, 174, 175, 176
Green-Wood Cemetery, 6, 57, 65–69, *66*
Grosbeak, Blue, 71, 217, 218, 224, 241
Grosbeak, Rose-breasted, 33, 60, 114, 204, 206, 227, 232
Gull, Bonaparte's, 74–75, 97, 147, 148, 153, 155, 183, 191, 218, 244, 261
Gull, Glaucous, 4, 74, 243–44, 247, 261
Gull, Greater Black-backed, 74
Gull, Herring, 33, 74
Gull, Iceland, 4, 27, 41, 74, 152–53, 185, 218, 243–44, 261

Gull, Laughing, 34, 71, 94, *96*, 114, 148, 188, 191, 217
Gull, Lesser Black-backed, 175, 217
Gull, Ring-billed, 33, 74
Gull, Western, 7

Harbor Herons Complex, 177
Harlem River, 36, 43
Harrier, Northern, 70–71, 92, 97, 100, 103, 126, 148, 151, 158, 182–83, 218, 227, 241, 244, 247, 249
Hashomomuck, 256
Hawk, Broad-winged, 91, 107
Hawk, Cooper's, 27, 60, 68, 91, 100, 103, 107, 136, 143, 148, 151, 174, 196, 218, 227, 247
Hawk, Red-shouldered, 34, 126, 148, 151, 158, 227, 241
Hawk, Red-tailed, 2, *18*, 26, 37, 40, 43, 51–53, 63–64, 67, 103, 107, 109, 114, 117, 136, 142–43, 151, 167
Hawk, Rough-legged, 148, 182, 241
Hawk, Sharp-shinned, 27, 34, 60, 68, 91, 100, 103, 114, 143, 151, 167, 174, 218, 227
hawk: fall watches for, 7, 27, 95, 173, 183, 213, 215, 218; nest cams, 56
Heckscher State Park, 237–38
Hempstead Harbor, 208
Hempstead Lake State Park, 178, 192–194, *193*
Hendrix Creek, 80–81
Heron, Great Blue, 26, 34, 63–64, 67–68, 83, 91, 106, *111*, 134, 136, 143, 150, 153, 188, 196, 203, 205, 217, 247
Heron, Green, 60, 83, 91, 106, 134, 143, *150*, 169, 196
Heron, Little Blue, 91, 147, 148, 150, 164–65, 169, 191
Heron, Tricolored, 91, 191
Highland Park, 118

High Line, 55
High Rock Park and Conservation Center, 164–66
Hither Hills State Park, 264
Hoffman Island, 5, 54
Hook Mountain, NJ, 7
Hook Pond, 265
Hudson River Park Greenway, 17, 45–48, 54
Hulse Landing Road, 239, 241
Hummingbird, Ruby-throated, 34, 60, 91, 134, 143
Husing Pond Preserve, 258

Ibis, Glossy, 71, *78*, 83, 91, 103, 148, 150, 182, 188, 191, 198, 219, 249
Important Bird Areas, 19, 57, 85, 121, 131, 138, 162, 180, 185, 192, 203, 215, 220, 222, 224–25, 228, 230, 235, 237, 248, 251, 254
Inlet Pond County Park, 256
Inwood Hill Park, 6, 17, 29–35, *30–31*

Jacob Riis Park, 85, *94–95*, 94–99
Jaeger, Parasitic, 217, 221
Jamaica Bay Wildlife Refuge, 5, 6, 85, 87–93, *88*; Big John's Pond, 89; East Pond, 87–90; West Pond, 90
Jay, Blue, *18*, 27, *119*, 205
John F. Kennedy Memorial Wildlife Sanctuary, 198
Jones Beach State Park, 6, 178, 180–184, *180–81*; Coast Guard Station, 181; dunes, 182; marina, 181; tip, 181; West End, 181; Zach's Bay, 181
Junco, Dark-eyed, 60, 62, 64, 68, 107, 109, 151, 205, 236, 262

Kestrel, American, 41, 43, 51, 68, 71, 148, 218

Killdeer, 37, 40–41, 76, 83, 91, 111, 114, 134, 147, 148, 158, 169, 174, 187–88, 217, 227, 238, *241*

Kingbird, Eastern, 26, 37, 40, 60, 64, *67*, 68, 112, 114, 134, 151, 153, 196, 217

Kingfisher, Belted, 34, 67, 68, 91, 106, 111, 114, 143, 150, 165–67, 169, 172, 257

King Fisher Park, 173–74

Kinglet, Golden-crowned, 2, 26, 34, 64, 92, 107, 114, 126, 158, *206*, 232

Kinglet, Ruby-crowned, 26, 34, 64, 92, 107, 114, 126, 148, 158, 211

Kissena Park and Corridor, 116–17

Kitttiwake, Black-legged, 221, 261

Knot, Red, 3, 91, 181–82, 247, 249–50

Lark, Horned, 41, 71, 76, 95, 97, 100, 115, 145, 147–48, 172, 181, 183, 209, 218, 238, 242

LaTourette Park, 176

Laurel Lake, 258

Leeds Pond Preserve, 206

Lemon Creek Park, 139, 154–55

Lofts Pond Park, 200

Long Pond Park, 158, 159

Longspur, Lapland, 148, 172, 183, 218

Loon, Common, 40–41, 71, 76–77, 97, 100, 103, 126, 148, 153, 155, 160, 183, 185, 209, 218, 234, 244, 246–47, 251, 257, 261

Loon, Red-throated, 41, 47, 71, 76, 97, 100, 126, 148, 153, 155, 160, 183, 185, 207, 218, 244, 247, 251, 261

MacKay Lake, 241

Madison Square Park, 50

Mallard, 3, 18, 34, 37, 47, 74, 109, 152, *163*, 196, 199, 227, 232, 247

Manhattan, *16*, 17–56

Marine Park. *See* Salt Marsh Nature Center

Mariner's Marsh Park, 166–168, *167*

Marratooka Lake Park (Marratooka Pond), 258

Martin, Purple, 139, 147–48, 153–54, *155*, 239

Mashomack Preserve on Shelter Island, 6, 215, 251–53, *252*

Massapequa Preserve, 178, 185–98, 203

Meadowlark, Eastern, 103, 217–18, 241

Mecox Bay, 250–51

Merganser, Common, 60, 91, 196, 199, 236

Merganser, Hooded, 24, 27, 60, 68, 74, 81, 91, 126, 134, 144, 150, 169, 175, 188, 196, 199, 234, 236

Merganser, Red-breasted 24, 34, 40, 44, 53, 60, 74, 76–77, 91, 100, 103, 118, 126, 148, 151, 153, 155, 160, 173, 199, 207, 209, 234, 244, 246, 249

Merlin, 34, 60, 71, 91, 97, 103, 148, 183, 217–18, 221, 227, 232

Midland Beach, 172–73

migration, 2–4, 8, 17, 19, 23, 53, 57, 178, 213

Milburn Pond, 200–201

Miller Field, 172–73

Mill Pond, 229–30

Mill Pond in Oyster Bay, 210

Mill Pond Park, 202

Mill Rock, 38–39, 41, 43

Mockingbird, Northern, *xvii*, 27, 108, 147

Montauk Peninsula, 258, *259*; Lake Montauk, 260; Lazy Point, 260–61; Montauk dump, 262; Napeague, 260

Montauk Point State Park, 6, 213, 258, 260

Moore's Woods, 256
Moriches Bay, 247
Moriches Inlet, 248
Morningside Park, 52
Moses, Robert, 180
Moses Mountain, 7, 139, 164–66, 174
Mount Loretto Unique Area, 6, 139, *156*, 156–59
Murre, Common, 7
Muscota Marsh, 17, 32, 35
Muttontown Preserve, 203–6, *204*

Napeague Bay, 264
Nassau County, 178–212, *179*; North Shore, 203; South Shore, 198; South Shore Winter Freshwater Birding, 6, 198–99
Nassau Point of Little Hog Neck, 257
National Gateway Recreational Area, Sandy Hook, NJ, 7
Nature Conservancy, 211, 212, 234, 242, 251, 253
nest cams, 56
New York Botanical Garden, 6, 121, 127–30; Bronx River, 129; Magnolia Way Road, 129; Mitsubishi Riverwalk, 129; Thain Family Forest, 129
New York City Audubon Society. *See* Audubon, New York City
New York Department of Environmental Protection, 8
New York Harbor boat tours, 5
Nickerson Beach, 6, 178, 186–87
Night-heron, Black-crowned, 26, 34, 37, 39, 67–68, 91, 103, 111–12, 114, 138, 143, 148, 150, 153, 167, 188, 202, 206, 209, 211, 231
Night-heron, Yellow-crowned, 91, 94, 103, 126, 138

Nissequogue River State Park, 231
North Brother and South Brother Islands, 5, 138
North Mount Loretto State Forest, 139, *156*, 156–159
North Woods, 19, 20, 24, *25*, 28
Nuthatch, Red-breasted, 25, 27, 34, 64, 67
Nuthatch, White-breasted, 27, 34, 67, 144
NYNYBirds, 28
NYSBirds-L, 7

Oakland Lake, 109–115, *110*
Oakwood Beach, 147, 148, 174
Oceanside Marine Nature Study Area, 6, 178, 189–92, *190*
Old Place Creek Park, 168–70
Olmsted, Frederick Law, 19, 25, 28, 52–53, 57, 85, 104, 208
Orient Beach State Park, 254–55
Orient Point County Park, 254
Oriole, Baltimore, 23, 27, 33–34, 37, 41, 60, 64, 68, 91, 106, 111, 114, 126, 129, 134, 143, 148, 150, 153, 160, 163, 167, 172, 176, 191, 196, 204, 208, 217, 226–27, 236–37, 252, 256, 261, 264
Oriole, Orchard, 26, 34, 60, 126, 153, 158, 167, 176, 217, 227, 252
Osprey, 3–4, 34, 71, *78*, 79, 83, 91, 94, 103, 114, 124–26, 143, 145, 148, 153, 155, 158, 160, 182, 188, 190–91, 196, 207–8, 217, 222, 225–27, 236–37, 239, 244, 247, 251–58, 261
Ovenbird, *23*, 68, 150, 227, 236, 242, 264
Owl, Barn, 90, 92
Owl, Barred, 126
Owl, Eastern Screech, 21, 27, 33, 150, 158, 196, 227, 235–36
Owl, Great Horned, 8, 27, 68, 114, 123,

126, 129, 158, 196–97, 210, 227, 234, 236, 252, 256

Owl, Long-eared, 123, 126, 183

Owl, Northern Saw-whet, 123–24, 126, 183

Owl, Short-eared, 182–83, 241

Owl, Snowy, 4, 70, 71, 148, 181, 183, 244

Owls Head Park, 75, 83

Oystercatchers, American, 2–3, 11, 71, 73, 83, 91, 95, 97, 100, 103, 116, 147, 148, 153, 173–74, 182–83, 186–87, 191, 216, 217, 244, 249

Pale Male, 26

Palisades Park, NJ, 7

Parakeet, Monk, 46, 65, 83

Parula, Northern, 44, 63, 67, 91, 97, 134, 227

Patchogue Lake, 223

Pelagic birds, 5, 7, 43, 73–74, 213, 217, 221, 243–44, 249, 260–61

Pelham Bay Park, 6, 7, 121–27, 122; Bartow-Pell Mansion, 121–22, 124, 126; Eastchester Bay, 125; Hunter Island, 123–24; Huntington Woods, 125; the Lagoon, 124; Meadow, 125; Orchard Beach, 122–23, 126; Turtle Cove, 124; Twin Island, 123

Peter Detmold Park, 53

Petrel, Fea's, 7

Pewee, Eastern Wood-, 26, 33, 60, 106, 112, 114, 143, 236, 242

Phalarope, Red-necked, 292

Pheasant, Ring-necked, 109, 148

Phoebe, Eastern, 2, 26, 60, 68, 91, 97, 107, 143, 148, 150, 160, 165

Pintail, Northern, 27, 91, 112, 114, 166–67, 169, 185, 191, 196, 199, 202, 230, 234

Pipit, American, 41, 71, 97, 183, 218

Planting Fields Arboretum State Historic Park, 210

Plover, American Golden, 182, 243, 251

Plover, Black-bellied, 84, 91, 148, 153, 183, 185, 188, 246–47, 249, 257

Plover, Piping, 2–3, 10, 11, 100, 103, 116, 182, 186–87, 217, 219, 221, 224–27, 232, 244, 249, 251, 254, 255, 257

Plover, Semipalmated, 217, 226–27

Plumb Beach, 77, 79, 83

Plum Beach. See Plumb Beach

Plum Island, 254

Point Breeze, 77

Point Lookout, 6, 178, 185–86

Ponquoge Bridge, 246

Prall's Island, 177

Preston's Pond, 242

Prospect Park, 6, 57, 58–62, 59; Boat House, 60; Long Meadow, 58; Lullwater, 60; Midwood, 58; Nellie's Lawn, 58; Nethermeade, 60; the Peninsula, 60; the Pools, 58; Prospect Lake, 60; Quaker Hill, 58; the Ravine, 60; Rose Garden, 58; Vale of Cashmere, 58

Puffin, Atlantic, 7

Queens, 85–119, 86

Queens Botanical Garden, 85, 108–9

Quogue Village Wetlands Preserve, 246

Quogue Wildlife Preserve, 246

Rail, Clapper, 91, 94, 125, 188, 191, 246–47, 249

Ramble, the, 19–22, 22, 28–29

Ram Island County Park, 251

Randall's Island, 6, 18, 38–43, 39

Raoul Wallenberg Forest Preserve, 138

raptors, 3, 4; in Bronx, 125–26, 138; in

Brooklyn, 69–70; in Manhattan, 41, 46; in Nassau County, 183, 188, 205, 209; in Queens, 95, 100, 102–3, 118; in Staten Island, 148, 151, 157–58, 175; in Suffolk County, 215, 218, 232, 238–39, 245, 249, 253, 255

Raritan Bay, 149, 150, 154, 156, 160, 167, 172

Razorbill, 5, 100, 100, 183, 244, 261

Redhead, 113–15, 118, 196, 199, 211, 230, 236

Redpoll, 4, 27, 181–83, 218

Redstart, American, 3, 37, 44, *63*, 97, 160, 165, 194, 205, 229

Reed's Basket Willow Swamp, 176

Ridgewood Reservoir, 106, 118

Riverdale Park, 138

Riverside Park, 52–53

Robert Moses State Park, 6, 7, 213, 215, *217*, 218; Democrat Point, 6, 217; Hawk Watch, 215, 218; Lighthouse, 216, 218

Robin, American, 112, 205

Rockaway Beach Endangered Species Nesting Area, 116

Rockaway Community Park. *See* Edgemere Landfill

Roosevelt, Theodore, 178, 211

Ruth Oliva Preserve at Dam Pond, 255–56

safety during birdwatching, 13–14

Sagamore Hill, 211

Salt Marsh Nature Center at Marine Park, 79

Sanderling, 4, 148, *152*, 183, *220*, 221, 226–27, 232, 247, 252

Sandpiper, Baird, 217

Sandpiper, Least, 32–34, 37, 71, 100, 148, 153, 169, 188, 217, 238, 244, 247

Sandpiper, Pectoral, 31, *32*, 33, 91, 169, 217, 238, 243, 249

Sandpiper, Purple, 97, 100, 148, 173–74, 183, 243–44, 260

Sandpiper, Semipalmated, 32–34, 37, 71, 91, 100, 169, 188, 217

Sandpiper, Solitary, 37, 91, 143, 169, 188, 237

Sandpiper, Spotted, 32–34, 37, 41, 60, 83–84, 91, 100, 114, 143, 148, 150, 153, 169, 207, 237

Sandpiper, Stilt, 91, 169, 191

Sandpiper, Upland, 217, 243

Sandpiper, Western, 191, 217

Sandpiper, White-rumped, 91, 153, 182, 191, 247, 249

Sands Point Preserve, 178, 207

Sapsucker, Yellow-bellied, 33, 39, 41, 51, 64, 107

scaup, *80*

Scaup, Greater, 34, 44, 71, 74, 91, 103, 158, 174, 247

Scaup, Lesser, 44, 71, 91, 112, 247

Scoter, Black, 183, 247, 261

Scoter, Surf, 183, 221, 261

Scoter, White-winged, 183, 261

Seaman Pond, 202

Shadmoor State Park, 263–64

Sharrotts Pond, 162, 163

Shearwater, Audubon's, 5

Shearwater, Cory's, 5, 217

Shearwater, Great, 5

Shearwater, Sooty, 217

Shelter Island, 251–53

Sherman Creek, 33, 36–38

Shinnecock Bay and Inlet, 6, 215, 243–45, *244*, 246

Shirley Marina County Park, 221

Shooters Island, 177

shorebirds, 3–4; in Brooklyn, 67, 73, 77,

79, 81–83; in Manhattan, 31, 32, 36–37, 41, 43–44; in Nassau County, 181–83, 185, 187–88, 191, 196, 198, 208–11; in Queens, 85, 87, 89–91, 95, 97, 99–100, 102; respect for nesting, 10–11; in Staten Island, 147–48, 152–53, 155, 159, 168–69; in Suffolk County, 213, 217, 219, 220, 225–27, 230–32, 234, 237–39, 241, 243–53, 255–57, *261*

Shoveler, Northern, 24, 27, 68, 91, 114–15, 196, 199

Shu Swamp (Charles T. Church Nature Sanctuary), 210

Silver Lake Park, 7, 175

Siskin, Pine, 91, 114, 126, 143, 181, 183, 218

Skimmer, Black, *v*, 3, 84, 91, 99, 100, 148, 155, 174, 182, 186–88, 191, 251

Smith Point County Park, 220–22

Snipe, Wilson's, 169, 196–197, *229*, 232

Snug Harbor, 139, 170–71

songbirds, 2–4, 8; in Bronx 124–25, 129, 134, 138; in Brooklyn, 62–64, 68, 71, 83–84; in Manhattan, 22, 27, 32, 36–37, 41, 44, 48, 51–52; in Nassau County, 181–83, 191–92, 194, 196, 202, 205–6, 208–10; in Queens, 87, 91, 95, 102–3, 109, 117; in Staten Island, 143, 148, 150, 153, 155, 158–60, 166–67, 175–76; in Suffolk County, 213, 215, 217, 222, 225, 227, 229–32, 234–38, 242, 246, 252, 254–58, 260–61, 264

Sora, 191

South Beach, 172–73, 175

South Shore Winter Freshwater Birding, 198–99

Sparrow, American Tree, 27, 60, 92, 125, 148, 158, 181, 191, 218, 232, 241, 247

Sparrow, Chipping, 27, 40, 44, 62, 63, 91, 108, 126, 143, 148, 160, 194, 204, 217, 226–27

Sparrow, Clay-colored, 97, 183, 216, 217

Sparrow, Field, 91, 97, 126, 143, 148, 217, 227, 236, 241, 252

Sparrow, Fox, 60, 92, 97, 107, 114, 126, 134, 148, 191, 196, 232

Sparrow, Grasshopper, 241

Sparrow, Ipswich, 218, 246–47

Sparrow, Nelson's, 40–41, 126

Sparrow, Saltmarsh, 41, 84, 188, 190–91

Sparrow, Savannah, 41, 60, 84, 91, 97, 103, 126, 134, 143, 148, 153, 160, 218, 227, 247

Sparrow, Seaside, 190–91

Sparrow, Song, 33, 37, 40–41, 43–44, 60, 64, 84, 91, 106–108, 114, 143, 153, 165, 169, 188, 204, 227, 241, 246–47, 262, *263*

Sparrow, Swamp, 34, 40–41, 60, 84, 91, 97, 114, 126, 134, 143, 165, 196, 217, 227

Sparrow, Vesper, 40, 97, 183, 218, 227

Sparrow, White-crowned, 27, 40–41, 60, 91, 97, 126, 143, 148, 183, *201*, 227, 241–42, 252

Sparrow, White-throated, 4, 27, 34, 40–41, 51, 60, 62, 64, 68, 92, 97, 107, 126, 144, 151, 153, 160, *171*, 188, 236

Spring Creek Park, 81–82

Spuyten Duyvil Shorefront Park, 33, 138

Staten Island, 139–177, *140*; Greenbelt, 176–77

Staten Island Ferry, 139, 177

Stehli Beach Preserve, 209

St. John's Pond Preserve, 211–12

Storm-petrel, Band-rumped, 7; Leach's, 7; Wilson's, 217, 249

Stump Pond, 234

Suffolk County, 213–65, *214*; central, 6, 239–43, *240*; North Fork, 253–65; Shelter Island, 251–53; South Fork, 243–51

Sunken Forest at Sailors Haven, 219
Sunken Meadow State Park, 215, 230–33, *231*
Swallow, Bank, 148, 153, 157–58, 193, 217, 226–27, 232, 234, 252, 254, 263
Swallow, Barn, 26, 33, 37, 40–41, 44, 63, 71, 91, 94, 97, 102–3, 106, 111–12, *113*, 114, 125, 134, 145, 148, 153, 169–70, 175, 182, 192–93, 196, 204, 217–18, 227, 244, 252
Swallow, Cliff, 193, 218
Swallow, Northern Rough-winged, 26, 33, 40, 44, 134, 148, 193, 196
Swallow, Tree, *ii*, 26, 33, 41, 44, 63, 71, 91, 94, 97, 103, 106, 111, 113, 125, 134, 148, 153, 175, 182, 193, 203–4, 217, 227, 244, 262
Swan, Mute, 112, 247
Swan Pond, 242
Swift, Chimney, 18, 26, 33, 41, 44, 60, 64, 114, 143, 207, 226–27
Swinburne Island, 5, 54
Swindler Cove, 33, 36–38

Tackapausha Museum and Preserve, 178, 195–98, 203
Tanager, Scarlet, 26, 113–14, 143, 165, 204, 227, 236, 256
Tanager, Summer, 26, 143, 172
Target Rock National Wildlife Refuge, 228–29
Teal, Blue-winged, 91, 188, 232
Teal, Green-winged, 71, 167, 169, 172, *183*, 188, 199, 232
Tern, Arctic, 249
Tern, Black, 182, 217, 244, 247, 249–50
Tern, Caspian, 4, 250, 261
Tern, Common, 18, 43–44, *45*, 47–48, 73, 83, 91, 97, 100, 103, 148, 186–88, 191, 219, 227, 244, 247, 254, 257, 261

Tern, Forster's, 91, 148, 153, 187–88, 191, 196, 247
Tern, Gull-billed, 91, 182
Tern, Least, 3, 10, 73, 91, 100, 103, 116, 148, 182, *186*, 187, 209, 227, 232, 251–52
Tern, Roseate, 100, 187, 217, 244, 247, 249, 254, 260–61
Tern, Royal, 4, 147, 148, 183, 217, 221, 247, 249
Terrapin, Diamondback, 91
Thrasher, Brown, 27, 33, *54*, 60, 91, 147–48, 167, 182, 217
Thrush, Bicknell's, 106
Thrush, Gray-cheeked, 27, 106
Thrush, Hermit, 27, 34, 64, 106, 109, 114, 126, 134, 136, 196
Thrush, Swainson's, 27, 68
Thrush, Wood, 27, 33, 60, 106, 112, 114, 151, 196, 204, 210, 227, 232
Tiana Beach, 246
Titmouse, Tufted, 27, 34, 39, 64, 205, 236
Tobay Beach, 198
Tottenville, 139, 175
Tottenville Train Station, 175
Towhee, Eastern, 27, 33, 63–64, 73, 91, 97, 114, 125, 151, 159–60, 165, 169, 172, 176, 182, 196, 207, 234, 236, 252, 257, 261
Tung Ting Pond, 229–30
Turkey, Wild, 126, 136, *137*, 222, 242
Turnstone, Ruddy, 91, 100, 148, 153, 207, 244, *245*, 246–47, 257
Twin Lakes Preserve, 202

Union Square Park, 50–51
Uplands Farm Sanctuary, 212

Van Cortlandt Park, 131–36, *133*
Vaux, Calvert, 19, 28, 57
Veery, 204, 227

Vireo, Blue-headed, 26, 33, 91, 106, 113, 165, 218
Vireo, Philadelphia, 26, 106
Vireo, Red-eyed, 26, 33, 44, 106, 114, 125, 150, 159, 165, 196, 204, 207, 218, 242, 256, 261, 264
Vireo, Warbling, 26–27, 33, 37, 60, 71, 106, 111–12, 114, 125, 134, 148, 153, 158, 172, 196, 204, 258
Vireo, White-eyed, 26, 60, 71, 91, 106, 113, 148, 158, 163, 204, 232, 236, 252
Vireo, Yellow-throated, 26, 106, 113

wading birds: in Bronx, 124–25; in Brooklyn, 58, 67, 71, 73, 79, 83–84; in Manhattan, 26, 36; in Nassau County, 188, 191, 198, 208, 224, 226, 234–36; in Queens, 90–91, 97, 103; in Staten Island, 143, 147, 149, 152–53, 160, 164, 169–72, 174, 176–77; in Suffolk County, 244, 248, 250, 255–57
Wading River Marsh Preserve, 238
Wantagh Pond, 202
Warbler, Bay-breasted, 60, 106, 113
Warbler, Black-and-white, xiv, 60, 63, 91, 125, 160
Warbler, Blackburnian, 24, 33, 91, 113, 167
Warbler, Blackpoll, 3, 33, 37, 44, 68, 113
Warbler, Black-throated Blue, 44, 63, 68, 91, 113, 144, 165, 167, 227
Warbler, Black-throated Green, 33, 60, 63, 91, 113
Warbler, Blue-winged, 33, 60, 106, 113, 212, 227, 234, 252, 264
Warbler, Cerulean, 106
Warbler, Chestnut-sided, 68, 113
Warbler, Common Yellowthroat, 37, 44, 63, 68, 91, 103, 148, 151, 158, 167, 169, 182, 196, 204, 227

Warbler, Hooded, 106
Warbler, Magnolia, 63, 91, 227
Warbler, Nashville, 33, 68
Warbler, Orange-crowned, 61
Warbler, Palm, 3, 33, 91, 123, 134, 160
Warbler, Pine, 33, 125, 134, 236, 242
Warbler, Prairie, 232, 241, 252, 257, 264
Warbler, Wilson's, 33, 91, 106
Warbler, Worm-eating, 106
Warbler, Yellow, 3, 37, 40, 41, 44, 60, 63, 68, 91, 103, 111, 114, 125, 134, 147–48, 160, 167, 169, 172, 182, 188, 194, 204, 227, 236, 242, 256–58
Warbler, Yellow-rumped, 3, 25, 44, 91, 160, 181, 183, 188, 232, 252
Warbler, Yellow-throated, 221
Washington Square Park, 51
Watch Hill, 219
waterfowl, 2, 4; in Bronx, 123–24, 129, 138; in Brooklyn, 60, 63–64, 70–71, 73–77, 79–81, 84; in Manhattan, 20, 24, 27, 34, 36–37, 39, 41, 46–47, 53–54; in Nassau County, 181, 183, 185, 188, 191, 194, 196, 198–203, 206–11; in Queens, 90–91, 94–95, 97, 99–100, 103, 114–15, 117–18; in Staten Island, 143–45, 147–48, 150–54, 157–58, 160, 165–66, 170, 172–75, 177; in Suffolk County, 213, 215, 217, 221, 223–24, 229–30, 234–38, 241, 243, 249–50, 252–58, 260–61, 263–64
Waterthrush, Louisiana, 63
Waterthrush, Northern, 63, 132
Wave Hill, 7, 137
Waxwing, Cedar, 27, 34, 91, 109, 112, 114, 126, 143, 158, 160, 188, 191, 196, 207, 229
Welwyn Preserve, 208–9
Wertheim National Wildlife Refuge, 222–24

Westhampton Dunes Overlook County
Park, 247, 249
Whimbrel, 257
Whip-poor-will, 162, 227
Whitney Lake. *See* Whitney Pond Park
Whitney Pond Park, 207–8
Wigeon, American, 68, 71, 74, 75, 81, 91,
103, 115, 126, 144, 174, 191, 199, 236
Wigeon, Eurasian, 75, 126, 175, 196, 199,
211, 230
Wild Bird Fund, 9
Wildlife Conservation Society. *See*
Bronx Zoo
Wildwood State Park, 239
Willet, 71, 84, 91, 94, 103, 187, 188, *191*,
217, 246–47
William Cullen Bryant Preserve, 208
William Floyd Estate, 222
Willowbrook Park, 171–72
Willow Lake, 118
Willow Pond, 178, 199–200
Wolfe's Pond Park, 139, 151–54
Woodcock, American, 2, *51*, 71, 91–92,
125, 148, 150

Woodlawn Cemetery, 121, 136
Woodpecker, Downy, 27, 34, 39, 41, 62,
64, 67, 92, 107, 112, 114, 134, 144, 151,
153, 160, 167, 188, 196, 218
Woodpecker, Hairy, 34, 92, 107, 112, 114,
134, 144, 151, 153, 160, 232
Woodpecker, Red-bellied, 34, 39, 41, 62,
64, 67, 92, 107, 112, 114, 134, 144, 151,
153, 160, 196
Woodpecker, Red-headed, 25, 27, 158,
227
World's Fair Marina, 118
Wren, Carolina, 27, 60, 64, 91, 106, 114,
143, 204, 208, 218, 226, 229, 236
Wren, House, 60, 91, 143, 165
Wren, Marsh, 91, 125, 169, 190–91
Wren, Winter, 107, 143, 197, 211

Yankee Pier, 43
yellowlegs, *15*
Yellowlegs, Greater, 37, 91, 125, 148,
169, 188, 191
Yellowlegs, Lesser, 91, 125, 169, 188,
191